Adaptive Filtering - Recent Advances and Practical Implementation

Edited by Wenping Cao and Qian Zhang

Published in London, United Kingdom

IntechOpen

Supporting open minds since 2005

Adaptive Filtering - Recent Advances and Practical Implementation
http://dx.doi.org/10.5772/intechopen.91562
Edited by Wenping Cao and Qian Zhang

Contributors
Zeeshan Ahmad, Tae-Hak Lee, Sang-Gyu Lee, Jean-Jacques Laurin, Ke Wu, Erick Ulin-Avila, Juan Ponce-Hernandez, Javaid Ahmad Reshi, Shahzad Ahmad Bhat, Bilal Ahmad Para, Eman Shawky Amer, Ashwin Yadav, Naren Naik, Peeyush Awasthi, Mudambi R Ananthasayanam

First published in London, United Kingdom, 2021 by IntechOpen
IntechOpen is the global imprint of INTECHOPEN LIMITED, registered in England and Wales, registration number: 11086078, 5 Princes Gate Court, London, SW7 2QJ, United Kingdom
Printed in Croatia

British Library Cataloguing-in-Publication Data
A catalogue record for this book is available from the British Library

Additional hard and PDF copies can be obtained from orders@intechopen.com

Adaptive Filtering - Recent Advances and Practical Implementation
Edited by Wenping Cao and Qian Zhang
p. cm.
Print ISBN 978-1-83962-377-6
Online ISBN 978-1-83962-378-3
eBook (PDF) ISBN 978-1-83962-379-0

For EU product safety concerns:
IN TECH d.o.o., Prolaz Marije Krucifikse Kozulić 3, 51000 Rijeka, Croatia, info@intechopen.com or visit our website at intechopen.com.

Meet the editors

Wenping Cao (FIET, SMIEEE) received a BEng in Electrical Engineering from Beijing Jiaotong University, China, in 1991, and a Ph.D. in Electrical Machines and Drives from the University of Nottingham, England, in 2004 Prof. Cao is Chair Professor of Electrical Engineering at Anhui University, Hefei City, China. He received a "Royal Society Wolfson Research Merit Award" in 2016, the "Dragon's Den Competition Award" from Queen's University Belfast, in 2014, and "Innovator of the Year Award" from Newcastle University, UK, in 2013. He has been an associate editor for *IEEE Transactions on Power Electronics*, *IEEE Transactions on Industry Application*, *IET Power Electronics*, and several international journals. His research interests include fault analysis and condition monitoring of electrical machines and power electronics.

Qian Zhang received a BS from Anhui University, China, in 2004, an MEng from the University of Science and Technology of China in 2007, and a Ph.D. from Anhui University in 2014. She is currently an associate professor in the School of Electrical Engineering and director of the Electrical Engineering Department at Anhui University. She received the "New Talent of Teaching Award" from Anhui Province in 2013. Her research interests include nonlinear control, servo system modeling, load forecasting, and fault diagnosis of distribution networks.

Contents

Preface

Adaptive filters are digital filters designed to change their coefficients as per operational conditions for advanced control purposes. They play an increasingly important role in industrial applications with functions such as adaptive control, noise cancellation, echo cancellation, channel equalization, image restoration, and seismic system identification.

This book captures current state-of-the-art technologies in signal processing, parameter estimation, advanced control, and big data applied to adaptive filtering. It is organized into three sections: "Fundamentals of Signal Processing and Adaptive Filters", "Kalman Filters and Analysis", and "Practical Implementations and Applications". Chapters within these sections explain the fundamentals, analyze phenomena and design considerations, and demonstrate cases and applications. It is a practical reference for academic researchers, postgraduate students, engineers, and practitioners in the field.

The editors wish to acknowledge all the chapter contributors and reviewers for their valuable inputs to the book. They are also thankful to Author Service Manager Miss Dolores Kuzelj at IntechOpen. Her remarks, encouragement, and conscientious attitude were key factors in making this book a reality.

Wenping Cao and Qian Zhang
Anhui University,
Hefei, China

Section 1

Fundamentals of Signal Processing and Adaptive Filters

Chapter 1

Fundamentals of Narrowband Array Signal Processing

Zeeshan Ahmad

Abstract

Array signal processing is an actively developing research area connected to the progress in optimization theory, and remains the key technological development that attracts prevalent attention in signal processing. This chapter provides an overview of the fundamental concepts and essential terminologies employed in narrowband array signal processing. We first develop a general signal model for narrowband adaptive arrays and discuss the beamforming operation. We next introduce the basic performance parameters of adaptive arrays and the second order statistics of the array data. We then formulate various optimal weigh vector solution criteria. Finally, we discuss various types of adaptive filtering algorithms. Besides, this chapter emphasizes the theory of narrowband array signal processing employed in narrowband beamforming and direction-of-arrival (DOA) estimation algorithms.

Keywords: Adaptive algorithms, Adaptive arrays, Array signal processing, Beamforming

1. Introduction

Array signal processing [1, 2] is an indispensable technique in signal processing with ubiquitous applications. The fundamental principles and techniques of array signal processing are applicable in various fields such as sonar, radar, and wireless communications etc. Antenna array processing manipulate and process each sensor output according to a certain algorithm to achieve better system performance than just a single antenna, and estimate the signal parameters from the data accumulated over the spatial aperture of an antenna array. [3, 4]. These parameters of interest include the signal content itself, their DOAs, and power. To get this information, the sensor array data is processed using statistical and adaptive signal processing techniques. These techniques include parameter estimation and adaptive filtering applied to array signal processing. Meanwhile, it also plays an important role in the multi-input multi-output (MIMO) communication system and a waveform diversity MIMO radar system, by improving its performance, reducing the clutter, and increasing the array resolution [1–4].

All in all, there are numerous potential advantages of array signal processing techniques, such as improved system capacity, signal bandwidth, the space division multiple access (SDMA), high signal-to-noise ratio (SNR), frequency reuse factor, side-lobe offsets or nulls, degree of freedoms, and the resolution of the antenna array [5]. In this chapter, we introduce the basic principle of array signal processing techniques to further understand its implementation process and applications. We begin by formulating the signal mathematical model used as a basis for discussing

IntechOpen

array signal processing in beamforming and direction-of-arrival (DOA) estimation algorithms. We also provide some introductory materials about beamforming techniques, performance analysis parameters, and a brief overview of some basic beamforming algorithms.

2. Adaptive array signal model

Since the real signal transmission environment is complex, so a strict mathematical model is the basis for adaptive beamforming and lays the groundwork for the discussion of beamforming algorithms. To simplify the analysis of the model, the signal source used in this chapter is a narrowband signal, that is, the bandwidth of the received array signal is much smaller than the carrier frequency of the signal, assuming that [6]:

a. Each array element is an ideal omnidirectional point source, and the inter-element spacing is less than or equal to half-a-wavelength.

b. The number of received signals is known, and less than the number of array elements.

c. The signal sources are assumed to be in the far-field so that the signals impinging on the array can be regarded as a plane wave;

d. The spacing between array elements are equal, i.e., evenly spaced array;

e. The noise is zero-mean Gaussian white noise, and uncorrelated with the signal source.

f. The effect of mutual coupling between array elements is assumed to be negligible, i.e., the different element receives the same signal amplitude.

Although the above assumptions are not valid for wideband signal source, the fundamental model used for them is very similar. Therefore, this chapter focuses on the mathematical model based on narrowband signal beamforming principle.

Adaptive antenna arrays may have different geometrical configurations. Different spatial distribution of array elements leads to different array configurations, such as linear arrays, circular arrays, rectangular arrays, and triangular arrays etc. [7, 8]. For an arbitrary array structure with M-elements as shown in **Figure 1**, θ and ϕ denote the elevation angle and the azimuth angle, respectively. Vector $\boldsymbol{a}$ and $\boldsymbol{p}_i$ respectively denote the direction vector of the signal and the coordinates of the $i-th$ array element. Since the signal received by each array element has a certain delay relative to the origin of the coordinates, the delay time [9] for the signal received at the $i-th$ array element is

$$\tau_i = \frac{\boldsymbol{a}^T \boldsymbol{p}_i}{c} \tag{1}$$

where c is the speed of light, and

$$\boldsymbol{a} = \begin{bmatrix} -\sin\theta\cos\phi \\ -\sin\theta\sin\phi \\ -\cos\theta \end{bmatrix}. \tag{2}$$

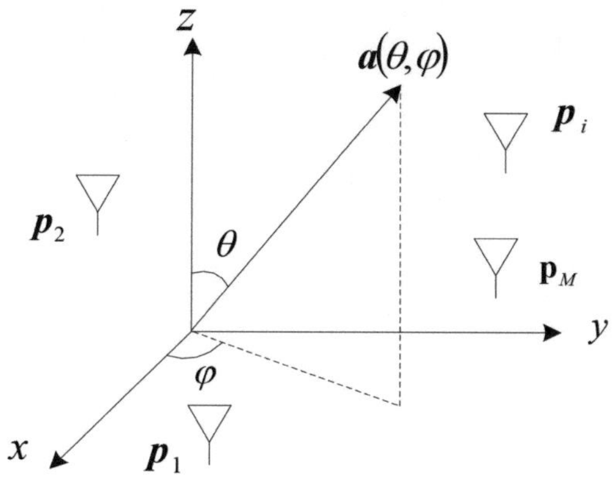

Figure 1.
Geometry of array.

$$\boldsymbol{p}_i = \begin{bmatrix} x_i \\ y_i \\ z_i \end{bmatrix}, \quad i = 1, 2, \cdots, M. \tag{3}$$

The signal received by the first sensor located at the origin of the coordinates is

$$\tilde{x}_1(t) = x_1(t)e^{j\omega t}. \tag{4}$$

The overall signal received by the array can be expressed as

$$\mathbf{x}(t) = \begin{bmatrix} x_1(t) \\ x_2(t) \\ \cdots \\ x_M(t) \end{bmatrix} = \begin{bmatrix} x_1(t-\tau_1)e^{j\omega(t-\tau_1)} \\ x_2(t-\tau_2)e^{j\omega(t-\tau_2)} \\ \cdots \\ x_M(t-\tau_M)e^{j\omega(t-\tau_L)} \end{bmatrix}. \tag{5}$$

If the received signal is a narrowband, we can ignore its amplitude changes for different elements. Consider the phase change only [10], the array received signal is simplified to

$$\mathbf{x}(t) = \begin{bmatrix} x_1(t) \\ x_2(t) \\ \cdots \\ x_M(t) \end{bmatrix} = x_1(t)\begin{bmatrix} e^{j\omega(t-\tau_1)} \\ e^{j\omega(t-\tau_2)} \\ \cdots \\ e^{j\omega(t-\tau_M)} \end{bmatrix} \tag{6}$$

Let us consider a uniform linear array (ULA) composed of M elements with inter-element spacing d as shown in **Figure 2**. Assume the first array element located at origin of coordinate as a reference element. Consider the far field source with P signals $s_0(t), s_1(t), \dots, s_{P-1}(t)$, having the same center carrier frequency f_c, the narrowband signal $s_i(t)$ impinges on the array at an angle θ_i relative to the broadside, which refers to the direction normal to the array, where $i = 0, \dots, P-1$ (without taking into account the azimuth angle, consider only the elevation angle).

Due to multipath propagation, each element receive the same signal with a different time delay. Due to the fact that the incident signal is a narrowband signal,

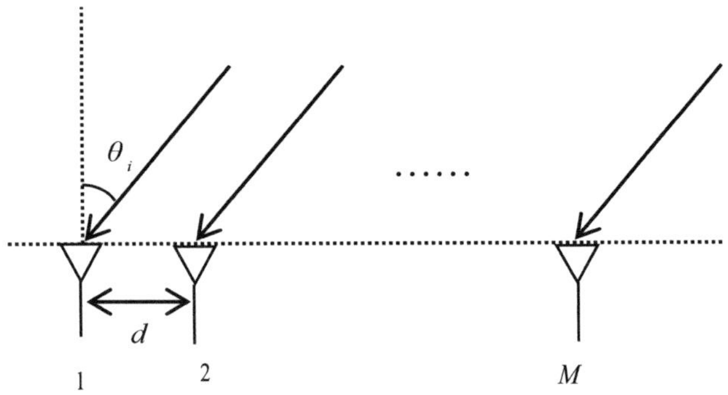

Figure 2.
Structure of uniform linear array antenna.

the amplitude variation is negligible, and only phase delay is considered. This delay is determined by the array element spacing d and the elevation angle of incidence. Consider the signal received by the first array element as a reference signal, then the analytical expression for the $i-th$ signal received with respect to the reference array element is

$$s_i(t) = m_i(t)e^{j2\pi f_c t}, \quad i = 0, \ldots, P-1 \tag{7}$$

where $m_i(t)$ is the complex envelope of the $i-th$ modulated signal, and f_c is the carrier frequency.

The propagation delay of the received signal from reference array element to the $m-th$ array element can be expressed as

$$\tau_m(\theta_i) = \frac{d}{c}(m-1)\sin\theta_i, \quad m = 1, \ldots, M. \tag{8}$$

According to Eq. (7), the signal received at the $m-th$ array element can be expressed as the superposition of all the signals, that is

$$x_m(t) = \sum_{i=0}^{P-1} m_i(t - \tau_m(\theta_i))e^{j2\pi f_c(t-\tau_m(\theta_i))} + n_m(t), \tag{9}$$

where $n_m(t)$ is the Gaussian noise signal received at the $m-th$ array element having zero mean and variance σ^2.

Since we consider a narrowband signal source located in the far-field, the bandwidth B of the signal satisfy the condition $B << f_c$, and $m_i(t)$ changes relatively slowly because the signal delay is $\tau_m(\theta_i) << \frac{1}{B}$. Therefore, complex envelope of the signal can be approximated as $m_i(t - \tau_m(\theta_i)) \approx m_i(t)$, that is, the difference in the array received signal complex envelope can be neglected. Thus, Eq. (9) is simplified as

$$x_m(t) = \sum_{i=0}^{P-1} m_i(t)e^{j2\pi f_c(t-\tau_m(\theta_i))} + n_m(t). \tag{10}$$

Since the carrier component in the system does not affect the analysis, and the adaptive algorithm is often carried out in the baseband (complex envelope), so the carrier part $e^{j2\pi f_c t}$ in the Eq. (10) can be ignored. Eq. (10) can then be expressed as

$$x_m(t) \approx \sum_{i=0}^{P-1} m_i(t) e^{-j(m-1)kd\sin\theta_i} + n_m(t). \tag{11}$$

where k is the free-space wave number given by [11].

$$k = 2\pi f_c / c. \tag{12}$$

At time t, the overall received signal can be expressed as

$$\boldsymbol{x}(t) = \begin{bmatrix} x_1(t) \\ x_2(t) \\ \vdots \\ x_M(t) \end{bmatrix} = \boldsymbol{Am}(t) + \boldsymbol{n}(t)$$

$$= \begin{bmatrix} 1 & 1 & \cdots & 1 \\ e^{-jkd\sin\theta_0} & e^{-jkd\sin\theta_1} & \cdots & e^{-jkd\sin\theta_{P-1}} \\ \vdots & \vdots & \vdots & \vdots \\ e^{-j(M-1)kd\sin\theta_0} & e^{-j(M-1)kd\sin\theta_1} & \cdots & e^{-j(M-1)kd\sin\theta_{P-1}} \end{bmatrix} \begin{bmatrix} m_1(t) \\ m_2(t) \\ \vdots \\ m_P(t) \end{bmatrix} + \begin{bmatrix} n_1(t) \\ n_2(t) \\ \vdots \\ n_M(t) \end{bmatrix} \tag{13}$$

where $\boldsymbol{A} = [\boldsymbol{a}(\theta_0) \ \boldsymbol{a}(\theta_1) \ \cdots \ \boldsymbol{a}(\theta_{P-1})]$ is the direction matrix (also called the array manifold matrix), $\boldsymbol{a}(\theta_i)$ is the direction vector for the $i-th$ signal $s_i(t)$, and $\boldsymbol{n}(t)$ is the noise vector, expressed as

$$\boldsymbol{a}(\theta_i) = \begin{bmatrix} 1 & e^{-j2\pi f_c \tau_1(\theta_i)} & \cdots & e^{-j2\pi f_c \tau_M(\theta_i)} \end{bmatrix}^T \tag{14}$$

$$\mathbf{n}(t) = [n_1(t) \quad n_2(t) \quad \cdots \quad n_M(t)]^T \tag{15}$$

where the sign $[]^T$ denotes the transpose operation.

3. Adaptive beamforming

Beamforming is a concept originating in array signal processing. The fundamental aim of beamforming is to estimate the desired signal properties by adjusting the complex weights at each sensor applied to the received signal which result in enhancement of desired signal and place nulls in the direction of interference. Adaptive arrays are capable to adjust its weights automatically according to the environment.

The beamforming can be classified into two types that are analog beamforming and digital beamforming [12].

The analog beamforming is performed in the analog domain. The block diagram of an analog beamformer is shown in **Figure 3**. The analog RF signal received by the antenna array is converted to an intermediate frequency by the RF front end, which is the analog intermediate frequency signal. The weight vector is calculated by the weights update algorithm. The weighted sum of the analog IF signal is obtained, and the array received signal is synthesized. At this point the signal is still analog signal; then by analog-to-digitical converter (A/D) the analog signal is sampled and quantized, and the analog IF signal is converted to a digital intermediate frequency signa. Then the digital IF signal is given to the next - level processing.

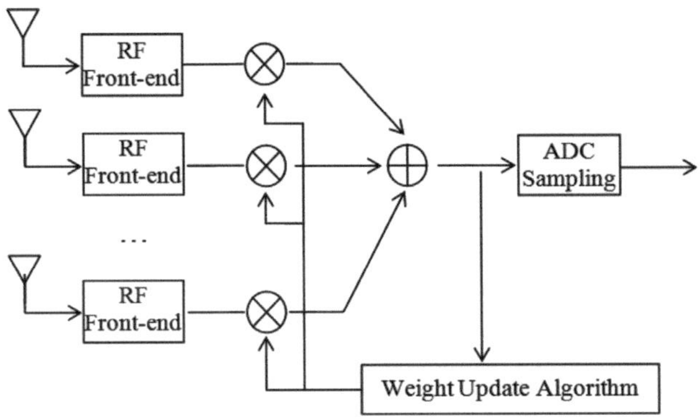

Figure 3.
The structure of analog beamforming.

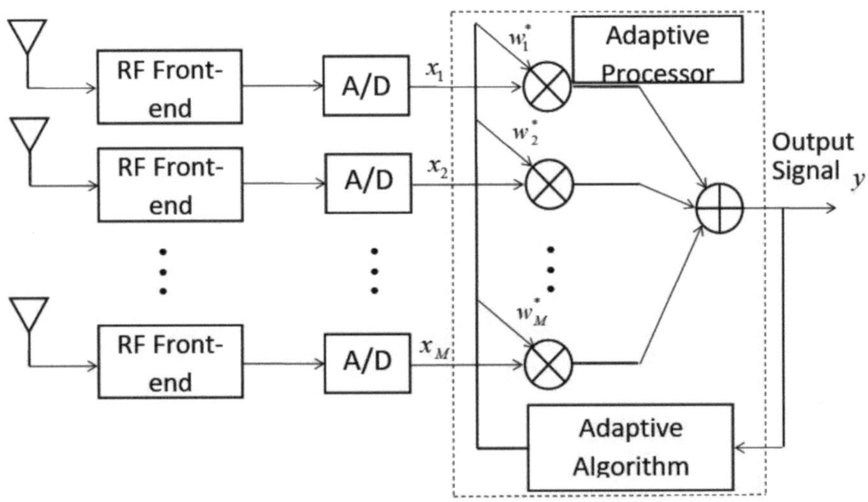

Figure 4.
Structure diagram of adaptive beamforming.

The digital beamforming is carried out in the digital domain, which is shown in the **Figure 4**.

Adaptive beamforming is a subclass of digital beamforming. Usually adaptive beamformer [13] comprises of RF Front-end, A/D converter module, and the signal processing (beam-control formation) module. A basic adaptive beamformer is shown in **Figure 4** which is composed of antenna array elements and an adaptive signal processor.

The antenna array elements receive the spatially-propagating desired signal and interference signal at the array aperture. In the RF Front-end, the received signal is down-converted to baseband signal [14], and then transformed into a digital signal through A/D converter, which is then processed by the adaptive processor. In adaptive processor, suitable adaptive filtering algorithm according to the requirements is applied to get the optimal weight vector. The weights are applied to the received signal at each array element to obtain a weighted sum of the signal. After the adaptive processing, the weighted signals are combined to get the output of the beamformer, which direct the main lobe in the direction of the desired signal and nulls in the directions of the interferers. The interference and noise are suppressed, and the output signal-to-interference-plus-noise ratio (SINR) of beamformer is thus improved.

Clearly, based on the adaptive beamformer structure shown in **Figure 4**, the output of each element is multiplied by a complex weight and summed to form the array output, $y(t)$, expressed as

$$y(t) = \mathbf{w}^H \mathbf{x}(t) = \sum_{m=1}^{M} w_m^* x_m(t) \tag{16}$$

where the symbol $[]^H$ represents the Hermitian (complex conjugate) transpose, $()^*$ indicates the conjugate, and $\mathbf{w}$ is the $M \times 1$ dimensional optimal weight vector computed by an adaptive filtering algorithm, given as

$$\mathbf{w} = \begin{bmatrix} w_1 & w_2 & \cdots & w_M \end{bmatrix}^T. \tag{17}$$

In this way, the array output, $y(t)$, is obtained by combining the weighted sum of each of the sensor signals. The different weight vectors for beamforming of signals from different directions have different response, thus pointing to the desired signal and suppress the interference signal.

Array output signal power is expressed as

$$P_{out} = E[y(t)^* y(t)] = \boldsymbol{w}^H \boldsymbol{R} \boldsymbol{w}. \tag{18}$$

where

$$\boldsymbol{R} = E\left[\boldsymbol{x}(t)\boldsymbol{x}^H(t)\right], \tag{19}$$

is the covariance matrix of the received signal, and $E[]$ denotes the expectation operator. Substitute Eq. (13) into Eq. (19), the covariance matrix can be expressed as

$$\mathbf{R} = \sum_{i=0}^{P-1} p_i \boldsymbol{a}(\theta_i) \boldsymbol{a}(\theta_i)^H + \sigma^2 \mathbf{I}, \tag{20}$$

where p_i is the power of signal $s_i(t)$, and $\mathbf{I}$ represents a identity/unit matrix. If the input signal in space has only one desired signal $s_0(t)$, and $P-1$ interference signals, then the covariance matrix can be expressed as

$$\begin{aligned} \mathbf{R} &= p_0 a(\theta_i) a(\theta_i)^H + \sum_{i=1}^{P-1} p_i a(\theta_i) a(\theta_i)^H + \sigma^2 \mathbf{I} \\ &= \boldsymbol{R}_s + \mathbf{R}_i + \mathbf{R}_n, \end{aligned} \tag{21}$$

where $\boldsymbol{R}_s$ is the covariance matrix of the desired signal, $\mathbf{R}_i$ is an interference signal covariance matrix, and $\mathbf{R}_n$ is the covariance matrix of the noise. Substitute Eq. (21) into Eq. (18), the output signal power can be expressed as a sum of desired signal power P_{os}, interference power as P_{oi} and noise power P_{on}.

$$P_{os} = \boldsymbol{w}^H \boldsymbol{R}_s \boldsymbol{w} \tag{22}$$

$$P_{oi} = \boldsymbol{w}^H \boldsymbol{R}_i \boldsymbol{w} \tag{23}$$

$$P_{on} = \boldsymbol{w}^H \boldsymbol{R}_n \boldsymbol{w} \tag{24}$$

The output SINR, a performance parameter of the beamformer, is defined as the ratio of the output desired signal power and the output power due to interference-plus-noise, and can be expressed as

$$SINR_{out} = \frac{P_{os}}{P_{oi} + P_{on}} = \frac{\mathbf{w}^H \mathbf{R}_s \mathbf{w}}{\mathbf{w}^H \mathbf{R}_i \mathbf{w} + \mathbf{w}^H \mathbf{R}_n \mathbf{w}}, \tag{25}$$

Adaptive antenna array takes the output SINR as an index to compute the optimal weights by maximizing the output SINR [15].

The most important performance indicator of the beamforming is the direction of the beampattern. It can be quite obvious to determine whether the resolution of any beamforming method is enough to enhance the desired signal and the extent of the suppression of interference signal is large enough. Array beampattern is defined as

$$B(\theta) = \left| \boldsymbol{w}^H \boldsymbol{a}(\theta) \right|. \tag{26}$$

When using analog beamforming, the hardware circuit is very complex, and the accuracy is low. In digital beamforming, the operations of phase shifting and amplitude scaling for each antenna element, and summation of received signals, are done digitally through a general-purpose DSP or dedicated beamforming chips. Therefore, digital beamforming is more flexible and do not require modification of the hardware structure.

Compared with analog beamforming, the digital beamforming has the following advantages:

a. Under the condition that the output SNR is not reduced and the hardware is not increased, digital beamforming can track multiple signals and form multi-beam.

b. The digital beamforming can make full use of the information received by the array antenna, real-time optimization of system performance, and achieve the real-time tracking of the desired signal.

c. In theory, digital beamforming can be achieved by implementing various algorithms.

d. Digital beamforming can achieve independent beamforming for each signal, and each beamforming can be optimized.

4. Basic parameters of adaptive array antenna

The performance parameters of an adaptive array antenna are basically the same as that of a single antenna, but because of the weight of the array, the specific values of each parameter depend on the array element characteristics, the weight vector, and geometry of the array [16].

4.1 Array pattern

The array pattern is the visual performance parameter of an antenna array. According to the pattern multiplication theorem of array antenna, the overall array pattern is the product of the element pattern $P_E(\varphi, \theta)$ and the array pattern $P_A(\varphi, \theta)$, that is

$$P(\varphi, \theta) = P_E(\varphi, \theta) P_A(\varphi, \theta). \tag{27}$$

Generally, it is assumed that the array elements are identical and omni-directional, hence

$$P_E(\varphi, \theta) = 1. \tag{28}$$

Thus, mostly adaptive array antenna patterns defined in the literature refers to the array factor part only, and the relationship between the received signal and the output signal is given as

$$y(t) = \boldsymbol{w}^{\mathrm{H}} \boldsymbol{x}(t). \tag{29}$$

Let's assume a single array element with the input signal power 1, the output signal power can be expressed as

$$P(\varphi, \theta) = \left| \boldsymbol{w}^{\mathrm{H}} \boldsymbol{a}(\varphi, \theta) \right|^2. \tag{30}$$

The above expression defines the power pattern of the array antenna. As can be seen from Eq. (30), the antenna beampattern is determined by the value of the weight vector; on the other hand, it also depends on the direction vector which is determined by the array geometry. Since we define the power pattern $P(\varphi, \theta)$ as the squared magnitude of the beampattern, therefore

$$B(\varphi, \theta) = \left| \boldsymbol{w}^{\mathrm{H}}(\theta_0) \boldsymbol{a}(\theta) \right|. \tag{31}$$

4.2 Array directivity and directivity index

The directivity of an adaptive array is closely related to the pattern of the array, which can be expressed as follows

$$D = \frac{4\pi P_{\max}(\varphi_0, \theta_0)}{\int_0^{\pi} d\theta \int_0^{2\pi} \sin\theta P(\varphi, \theta) d\varphi}, \tag{32}$$

where $P_{\max}(\varphi_0, \theta_0)$ is the maximum pattern that points to the direction of the main lobe.

The directivity is usually expressed in dB and is called array directivity index (DI) given by

$$DI = 10 \log_{10} D. \tag{33}$$

4.3 Array gain

The purpose of antenna array is to improve the G/T (gain of an antenna divided by its system temperature) ratio of the antenna. Array gain G is the main parameter to measure the SNR of the array, which is defined as the ratio of the output signal to noise ratio SNR_o and the input signal to noise ratio SNR_i.

$$G = \frac{SNR_o}{SNR_i}. \tag{34}$$

4.4 Sensitivity

The array beampattern is a function of weight vector and direction vector. However, due to the influence of various errors, the weight vector and the direction vector will have some errors, such as sensor position errors, covariance matrix estimation errors, inconsistent channel errors, and the mutual coupling between the array elements cause weight vector errors. Suppose the error-free weight vector $\boldsymbol{w}^0$ of the $m-th$ element is

$$\mathbf{w}_m^0 = g_m^0 e^{j\varphi_m^0}. \tag{35}$$

The $m-th$ element weight vector with error is

$$\mathbf{w}_m = \left(g_m^0 + \Delta g_m\right) e^{j\left(\varphi_m^0 + \Delta\varphi_m\right)}, \tag{36}$$

where the error Δg_m and $\Delta\varphi_m$ are zero mean Gauss random variables, and the variance is

$$Var(\Delta g_m) = \sigma_g^2 \tag{37}$$

$$Var(\Delta\varphi_m) = \sigma_\varphi^2 \tag{38}$$

For the direction vector, the error is mainly derived from the array element position errors. For the $m-th$ element, if there is no error in the array element position coordinates, then

$$\mathbf{p}_m^0 = \left[p_{mx} \quad p_{my} \quad p_{mz}\right]^{\mathrm{T}}. \tag{39}$$

While the coordinate with the error can be expressed as

$$\mathbf{p}_m = \left[p_{mx} + \Delta p_{mx} \quad p_{my} + \Delta p_{my} \quad p_{mz} + \Delta p_{mz}\right]^{\mathrm{T}}, \tag{40}$$

where the error quantity is Gauss random variable, which are zero mean, and the variance is

$$Var(\Delta p_{mx}) = Var\left(\Delta p_{my}\right) = Var(\Delta p_{mz}) = \sigma_p^2. \tag{41}$$

The array pattern at this instant is

$$P(\varphi, \theta) = P^0 e^{-\left(\sigma_\varphi^2 + \sigma_\lambda^2\right)} + \sum_{m=1}^{M} \left(g_m^0\right)^2 \left(1 + \sigma_g^2 - e^{-\left(\sigma_\varphi^2 + \sigma_\lambda^2\right)}\right), \tag{42}$$

where λ is the wavelength, and P^0 denotes the error-free pattern given by

$$P^0 = \left|\mathbf{w}^0 \mathbf{a}^0\right|^2, \tag{43}$$

and the variance is

$$\sigma_\lambda^2 = \left(\frac{2\pi}{\lambda}\right)^2 \sigma_p^2. \tag{44}$$

From Eq. (42), it is seen that the actual pattern consists of two parts. The first part is the error free pattern, i.e., the first term of the equation, and the error in the second term. In the second term, the coefficient g_m^0 is used to amplify the error, so the sensitivity of the array is defined as

$$T_s = \sum_{m=1}^{M} \left(g_m^0\right)^2. \tag{45}$$

5. Optimal beamforming

In beamforming, the weight vector is computed by solving the optimization of the cost function. The different cost functions corresponds to different criteria. Some of the most frequently used performance criteria's include minimum mean squared error (MMSE), maximum signal-to-interference-and noise ratio (MSINR), maximum likelihood (ML), minimum noise variance (MV), minimum output power (MP), and maximum gain, etc. [17]. These criteria's are often expressed as cost functions which are typically inversely associated with the quality of the signal at the array output. As the weights are iteratively adjusted, the cost function becomes smaller and smaller. When the cost function is minimized, the performance criterion is met and the algorithm is said to have converged.

5.1 Maximum signal-to-interferer-noise ratio

As can be seen from Eq. (21), the array output signal power consists of the desired signal power, interference power and noise power, and they are mutually uncorrelated. Since the interference signal and the noise is independent i.e. mutually uncorrelated and zero mean, so, $\mathbf{R}_{\mathrm{i}} + \mathbf{R}_n$ is a full rank and Hermite positive definite matrix. By unitary transformation it can be converted into unitary matrix as

$$\begin{aligned} \mathbf{U}^*(\mathbf{R}_{\mathrm{i}} + \mathbf{R}_n)\mathbf{U}^T &= \mathbf{U}^* E\left[\left(\sum_{i=0}^{P-1} m_i(t)\boldsymbol{a}(\theta_i)\right)\left(\sum_{i=0}^{P-1} m_i(t)a(\theta_i)\right)^H\right]\mathbf{U}^T + \sigma^2\mathbf{I} \\ &= E\left[\left(\mathbf{U}\sum_{i=0}^{P-1} m_i(t)\boldsymbol{a}(\theta_i)\right)^*\left(\mathbf{U}\sum_{i=0}^{P-1} m_i(t)\boldsymbol{a}(\theta_i)\right)^T\right] + \sigma^2\mathbf{I} \\ &= \sigma^2\mathbf{I} \end{aligned} \tag{46}$$

If we make

$$\boldsymbol{w} = \boldsymbol{U}^T\hat{\boldsymbol{w}}, \tag{47}$$

the output SINR will be

$$\begin{aligned} SINR_{out} &= \frac{\hat{\mathbf{w}}^H E\left[(\mathbf{U}m_0\boldsymbol{a}(\theta_0))^*(\mathbf{U}m_0\boldsymbol{a}(\theta_0))^T\right]\hat{\mathbf{w}}}{\hat{\mathbf{w}}^H E\left[\left(\mathbf{U}\sum_{\mathrm{i}=0}^{P-1} m_i\boldsymbol{a}(\theta_i)\right)^*\left(\mathbf{U}\sum_{\mathrm{i}=0}^{P-1} m_i\boldsymbol{a}(\theta_i)\right)^T\right]\hat{\mathbf{w}} + \sigma^2\mathbf{I}} \\ &= \frac{\hat{\mathbf{w}}^H E\left[(m_0\mathbf{U}\boldsymbol{a}(\theta_0))^*(m_0\mathbf{U}`\boldsymbol{a}(\theta_0))^T\right]\hat{\mathbf{w}}}{\|\hat{\mathbf{w}}\|^2}. \end{aligned} \tag{48}$$

According to Cauchy-Schwartz inequality

$$SINR_o \leq \|m_0 \mathbf{U} \boldsymbol{a}(\theta_0)\|^2 = E\left[|m_0|^2\right] \cdot \|\mathbf{U}^* \boldsymbol{a}(\theta_0)\|^2. \tag{49}$$

When the equality holds, then

$$\hat{\mathbf{w}} = \mathbf{U}^* \boldsymbol{a}(\theta_0). \tag{50}$$

The optimal solution for the weight vector

$$\mathbf{w}_{MSINR} = \mathbf{U}^T \mathbf{U}^* \boldsymbol{a}(\theta_0) = (\mathbf{R}_i + \mathbf{R}_n)^{-1} \boldsymbol{a}(\theta_0). \tag{51}$$

The optimal weight vector solution of the MSINR has the following advantages: only the DOA of the desired signal is required, and the DOA information for the interference signals is not needed; $\boldsymbol{R}_i + \boldsymbol{R}_n$ can be obtained through sampling and estimating the signal of each array element when the desired signal is interrupted; taking into account the constraints of the interference and noise signal, the output has a maximum SINR.

5.2 Minimum mean square error

Mean squared error refers to the mean squared difference between the beamformer output and the desired signal. The MMSE algorithm minimizes the error with respect to a reference signal $d(t)$. If the signal prior knowledge is known, the receiver can generate a local reference signal which has a strong correlation with the desired signal. The main idea of MMSE is to adjust the weight vector in real time, so that the mean squared error between the array output signal and the reference signal can be minimized. The estimator is of the form

$$y = \boldsymbol{w}^H \boldsymbol{x}. \tag{52}$$

The cost function, i.e., the mean square value of the error signal is

$$J(\boldsymbol{w}) = E\left[\left|\boldsymbol{w}^H \boldsymbol{x} - d\right|^2\right]. \tag{53}$$

Expanding the right-side of Eq. (53) and $\boldsymbol{w}$ should be taken out of the expectation operator, $E[\cdot]$, because it is not a statistical variable, we get

$$J(\boldsymbol{w}) = \boldsymbol{w}^H E\left[\boldsymbol{x}\boldsymbol{x}^H\right]\boldsymbol{w} - E\left[d\boldsymbol{x}^H\right]\boldsymbol{w} - \boldsymbol{w}^H E[\boldsymbol{x}d^*] + E[dd^*]. \tag{54}$$

According to the Lagrange multiplier method, in order to minimize the mean squared error function, taking the derivative with respect to $\boldsymbol{w}$ of the above expression

$$\begin{aligned} \frac{\partial}{\partial \boldsymbol{w}} J(\boldsymbol{w}) &= 2E\left[\boldsymbol{x}\boldsymbol{x}^H\right]\boldsymbol{w} - 2E[\boldsymbol{x}d^*] \\ &= 2\boldsymbol{R}\,\boldsymbol{w} - 2\boldsymbol{r}_{xd}, \end{aligned} \tag{55}$$

where $\boldsymbol{r}_{xd}$ is the cross-correlation vector between the input signal and the reference signal. Set the above result equal to 0 and solve for $\boldsymbol{w}$, the optimal MMSE weights are

$$\boldsymbol{w}_{MMSE} = \boldsymbol{R}^{-1} \boldsymbol{r}_{xd}. \tag{56}$$

Since the reference signal is only related to the desired signal, and is not related to the interference signal and noise, therefore

$$\mathbf{r}_{xd} = E[\mathbf{x}d^*] = E[m_0\boldsymbol{a}(\theta_0)d^*] = E[m_0 d^*]\boldsymbol{a}(\theta_0) = p_0\boldsymbol{a}(\theta_0), \tag{57}$$

and according to the matrix inversion formula

$$\mathbf{R}^{-1} = \frac{(\mathbf{R}_i + \mathbf{R}_n)^{-1}}{1 + p_0\boldsymbol{a}(\theta_0)^H(\mathbf{R}_i + \mathbf{R}_n)^{-1}\boldsymbol{a}(\theta_0)}. \tag{58}$$

Substitute Eq. (57) and Eq. (58) into Eq. (56), we get

$$\mathbf{w}_{MMSE} = \frac{p_0}{1 + p_0\boldsymbol{a}(\theta_0)^H(\mathbf{R}_i + \mathbf{R}_n)^{-1}\boldsymbol{a}(\theta_0)}\mathbf{w}_{MSINR}. \tag{59}$$

From the above analysis, it can be seen that the received signal is correlated with the desired signal. Therefore, it is not required to decompose the received signal into the desired signal and interference signal, and the correlation of the received signal and the reference signal can be estimated by sampling, so it is not difficult to determine.

On the other hand, from Eq. (59) it can be shown that the MMSE beamformer $\mathbf{w}_{MMSE}$ is a scalar multiple of the Max-SINR beamformer $\mathbf{w}_{MSINR}$ in Eq. (51), i.e., the adaptive weights obtained by using the MMSE and Max-SINR criteria are proportional to each other. Since the multiplicative constants in adaptive weights do not matter, these two techniques are therefore equivalent.

5.3 Minimum variance

In the signal received by the array, the desired signal is the content of cooperative communication, and the interference is often unpredictable, so the form of the desired signal and DOA of the signal should be known. In this case, in order to detect the desired signal more efficiently, it is necessary to eliminate the clutter background. From Eq. (22)-(24) it is shown that the array output power includes three parts: desired signal power, interference power and noise power, while the interference and noise power can be considered as the variance of the desired signal error. The smaller the variance is, the more close is it to the expectation. Interference and noise power can be expressed as

$$P_{oi} + P_{on} = \boldsymbol{w}^H\boldsymbol{R}_i\boldsymbol{w} + \boldsymbol{w}^H\boldsymbol{R}_n\boldsymbol{w} \tag{60}$$

For array main-lobe (desired look direction), the unit gain is considered, that is

$$\begin{cases} \min\limits_{w} & \boldsymbol{w}^H\boldsymbol{R}_{i+n}\boldsymbol{w} \\ s.t. & \boldsymbol{w}^H\boldsymbol{a}(\theta_0) = 1 \end{cases} \tag{61}$$

Therefore, the minimum interference and noise variance is the choice of the appropriate **w**, using the Eq. (61) constraints, so that the Eq. (60) is minimized. The weight vector **w** that minimizes Eq. (60) subject to the constraint in Eq. (61) can be selected by using a vector Lagrange multiplier to form the modified performance measure. According to Lagrange multiplier method, the objective function is

$$L(\mathbf{w}) = \mathbf{w}^H\mathbf{R}_i\mathbf{w} + \mathbf{w}^H\mathbf{R}_n\mathbf{w} + \lambda(\mathbf{w}^H\boldsymbol{a}(\theta_0) - 1) \tag{62}$$

Setting the derivative of the above expression Eq. (62) with respect to $\mathbf{w}$ equal to zero to obtain optimal weight vector $\mathbf{w}_{MV}$based on minimum variance criteria, requiring $\mathbf{w}_{MV}$ to satisfy the constraint in Eq. (61) to evaluate μ, and substituting the resulting value of μ into $\mathbf{w}_{MV}$ gives the minimum variance weight vector solution

$$\frac{\partial}{\partial \mathbf{w}} L(\mathbf{w}) = 2(\mathbf{R}_i + \mathbf{R}_n)\mathbf{w} + \lambda \boldsymbol{a}(\theta_0) = 0 \tag{63}$$

Solution of the above equation yields the optimal weights vector by the minimum interference and the noise variance criterion.

$$\mathbf{w}_{MV} = \mu(\mathbf{R}_i + \mathbf{R}_n)^{-1} \boldsymbol{a}(\theta_0) = \mu \mathbf{w}_{MSINR} \tag{64}$$

According to the constraint conditions of the main beam, using the property that $(\mathbf{R}_i + \mathbf{R}_n)$ is the Hermitian matrix, can be obtained as

$$\mu = \frac{1}{\boldsymbol{a}^H(\theta_0)(\mathbf{R}_i + \mathbf{R}_n)^{-1} \boldsymbol{a}(\theta_0)} \tag{65}$$

When the snapshot data used to estimate $\mathbf{R}$ contains only the noise and interference environment, this processor is referred to as minimum variance distortionless response (MVDR). In the event, the desired signal is also present in the snapshot data, the same solution for the weight vector results, but is sometimes referred to as minimum power distortionless response (MPDR) to indicate the difference in the observed data [2]. In practice, the distinction makes a significant difference in terms of the required snapshot support to achieve good performance [18].

5.4 Minimum power

The formulation of the MV can be derived by minimizing the total output power of the array subject to the similar constraint of distortion-less response of Eq. (61). The total power of the output signal is considered, if the gain of the desired signal is kept fixed, that is the same as the constraint condition of Eq. (61), which is equivalent to the received power of the signal under the condition of ensuring the normal receiving of the desired signal while suppressing interference and noise power, the resultant criterion is defined as the minimum total output power of the array (MP). The cost function is

$$\begin{cases} \min\limits_{w} \quad \boldsymbol{w}^H \boldsymbol{R} \boldsymbol{w} \\ s.t. \quad \boldsymbol{w}^H \boldsymbol{a}(\theta_0) = 1 \end{cases} \tag{66}$$

Also using the method of Lagrange multiplier, the objective function to be minimized is

$$L(\boldsymbol{w}) = \boldsymbol{w}^H \boldsymbol{R} \boldsymbol{w} + \lambda\left(\boldsymbol{w}^H \boldsymbol{a}(\theta_0) - 1\right) \tag{67}$$

Taking the complex gradient with respect to $\boldsymbol{w}$ and setting to zero

$$\frac{\partial}{\partial \boldsymbol{w}} L(\boldsymbol{w}) = 2\boldsymbol{R}\boldsymbol{w} + \lambda \boldsymbol{a}(\theta_0) = 0 \tag{68}$$

Under this criterion, the optimal weight vector is

$$\boldsymbol{w}_{MP} = \kappa \boldsymbol{R}^{-1}\boldsymbol{a}(\theta_0) \tag{69}$$

where the constant (normalize the array main beam gain to unity) is

$$\kappa = \frac{1}{\boldsymbol{a}^H(\theta_0)\boldsymbol{R}^{-1}\boldsymbol{a}(\theta_0)} \tag{70}$$

This criterion (MP) compared with the previously defined criterion (MV) is almost equivalent, since minimizing the total output power of the beamformer while preserving the desired signal is equivalent to minimizing the output power due to interference-plus-noise. The difference is only in the optimal weight vector of the MP criterion, and it is not necessary to separate the interference and noise, and only the covariance matrix of the received signal is estimated and thus the two optimization problems in Eq. (61) and Eq. (66) are equivalent.

5.5 Maximum likelihood criterion

Assume the space has only one desired signal and number of interference signals, the input signals can be expressed as

$$\boldsymbol{x} = m_0\boldsymbol{a}_0 + \sum_{i=1}^{M} m_i\boldsymbol{a}_i + \boldsymbol{n} = m_0\boldsymbol{a}_0 + \left(\sum_{i=1}^{M} m_i\boldsymbol{a}_i + \boldsymbol{n}\right) \tag{71}$$

If the interference signal and noise are zero mean Gaussian random process, the above equation is a Gaussian random process, and its mean is the desired signal $m_0\mathbf{a}_0$. The output signal is defined as the likelihood function vector

$$L(\boldsymbol{x}) = -\ln\left(P\left(\boldsymbol{x}\middle|\boldsymbol{x} = \sum_{i=1}^{M} m_i\boldsymbol{a}_i + \boldsymbol{n}\right)\right) \tag{72}$$

The expression of the conditional probability can be further changed to

$$L(\boldsymbol{x}) = c(\boldsymbol{x} - m_0\boldsymbol{a}_0)^{\mathrm{H}}(\boldsymbol{R}_i + \boldsymbol{R}_n)^{-1}(\boldsymbol{x} - m_0\boldsymbol{a}_0) \tag{73}$$

where c is a constant independent of $\boldsymbol{x}$ and $m_0\boldsymbol{a}_0$. Taking derivative of the above expression with respect to m_0 and set the result equal to zero, we will get the maximum likelihood estimation m_0

$$\frac{\partial}{\partial m_0}L(\boldsymbol{x}) = -2\boldsymbol{a}_0^{\mathrm{H}}(\boldsymbol{R}_i + \boldsymbol{R}_n)^{-1}\boldsymbol{x} + 2m_0\boldsymbol{a}_0^{\mathrm{H}}(\boldsymbol{R}_i + \boldsymbol{R}_n)^{-1}\boldsymbol{a}_0 = 0 \tag{74}$$

$$m_{0ML}(t) = \frac{\boldsymbol{a}_0^{\mathrm{H}}(\boldsymbol{R}_i + \boldsymbol{R}_n)^{-1}}{\boldsymbol{a}_0^{\mathrm{H}}(\boldsymbol{R}_i + \boldsymbol{R}_n)^{-1}\boldsymbol{a}_0}\boldsymbol{x} \tag{75}$$

The optimal weight vector is obtained by the above equation of the maximum likelihood criterion.

$$\boldsymbol{w}_{ML} = \frac{(\boldsymbol{R}_i + \boldsymbol{R}_n)^{-1}\boldsymbol{a}_0^{\mathrm{H}}}{\boldsymbol{a}_0^{\mathrm{H}}(\boldsymbol{R}_i + \boldsymbol{R}_n)^{-1}\boldsymbol{a}_0} \tag{76}$$

Compared with the weight vector solution under the Maximum Signal-to--Interferer-Noise Ratio (MSINR) criterion, the above expression can be rewritten as

$$w_{ML} = \frac{1}{a_0^H (R_i + R_n)^{-1} a_0} w_{MSINR} \tag{77}$$

From Eq. (77) it is clear that, the ML beamformer $\mathbf{w}_{ML}$ is a scalar multiple of the Max-SINR beamformer $\mathbf{w}_{MSINR}$ in Eq. (51). i.e., the adaptive weights obtained using the ML and Max SINR criteria are proportional to each other. Since multiplicative constants in the adaptive weights have no impact on the array beampattern, these two techniques have no essential difference and are therefore equivalent.

6. Adaptive filtering algorithms

The expression of the optimal weight vector is obtained by solving the equations based on the optimization theory. In practical engineering, the optimal weight vector is obtained by the adaptive filtering algorithms. When there is a reference signal available, the reference signal may be the training sequence of the desired signal or the DOA information of the desired signal, the resultant technique is categorized as a non-blind adaptive spatial filtering. These classical adaptive algorithms include Direct Matrix Inversion (DMI) [19], Least Mean Square (LMS) [20–22], Recursive Least Square (RLS) [23–25], Conjugate Gradient (CG) and its improved algorithms [26, 27]. When there is no reference signal available, the optimal weight vector solution can be obtained by using other characteristics of the signal, the resultant techniques are categorized as blind adaptive spatial filtering. Blind algorithm mainly includes Constant Modulus (CM) algorithm [28–30], smooth circulation (Cyclo-stationary) algorithm [31], and High Order Cumulant (HOC) algorithm [32].

6.1 Direct matrix inversion algorithm

The basic idea of DMI algorithm is to compute the optimal weight vector directly instead of calculating it iteratively, based on an estimate of the correlation matrix $R = E[x(t)x^H(t)]$ of the adaptive array output samples [33]. In communication systems, the signal source consists of a desired signal, interference and noise, therefore, the maximum SINR criterion, the minimum mean square error (MMSE) criterion, the minimum variance (MV) criterion and the maximum likelihood (ML) criterion need to know the covariance matrix of the interference signal and the noise signal, and do not contain the covariance matrix of the desired signal. So these criteria are not suitable for communication systems, and are suitable for radar systems, because it is easy to realize the interference and noise superimposed signal as long as the radar does not transmit the signal but only receives the signal.

For the MP criterion, the solution also needs the desired signal DOA, which is based on Eqs. (68) and (69), thus obtaining the desired signal direction vector $a(\theta_0)$. On the other hand, unlike the MV criterion, the signal covariance matrix of MP criterion is the sum of the covariance matrices of the desired signal, the interference and the noise. Therefore, the MP criterion is suitable for the communication system.

Assume that there are P signals in the space, wherein, the desired signal is $\mathbf{s}_0 = m_0 a(\theta_0)$, the power is p_0, and the interference signals are $\mathbf{s}_1 = m_1 a(\theta_1)$, ..., $\mathbf{s}_P = m_P a(\theta_{P-1})$ with power $p_1, \ldots, p_{P-1}$, respectively. The noise vector is n, and power is σ^2. According to the definition of covariance matrix

$$\begin{aligned}\mathbf{R} &= E\left[\left(\sum_{i=0}^{P-1}\mathbf{s}_i+\mathbf{n}\right)\left(\sum_{i=0}^{P-1}\mathbf{s}_i+\mathbf{n}\right)^H\right]\\ &= E\left[\left(\sum_{i=0}^{P-1}m_i\boldsymbol{a}(\theta_i)+\mathbf{n}\right)\left(\sum_{i=0}^{P-1}m_i\boldsymbol{a}(\theta_i)+\mathbf{n}\right)^H\right]\end{aligned} \tag{78}$$

Because the spatial separation between signal and interference is large enough, they are spatially uncorrelated. When sources are uncorrelated

$$E\left[\boldsymbol{a}(\theta_i)\boldsymbol{a}(\theta_j)^H\right]=0 \quad i\neq j \tag{79}$$

At the same time

$$E\left[m_i^2\boldsymbol{a}(\theta_i)\boldsymbol{a}(\theta_i)^H\right]=p_i \tag{80}$$

$$E\left[\boldsymbol{n}\boldsymbol{n}^H\right]=\sigma^2 \tag{81}$$

Obviously, in practical applications, it is very difficult to estimate the covariance matrix by the respective amount of power, instead it can be estimated from samples of the received signal. DMI algorithm assumes that the covariance matrix has been estimated, and the expression $\mathbf{R}^{-1}$ is obtained by matrix inversion, combine with the known DOA, calculate the direction vector $\boldsymbol{a}(\theta_0)$, and the optimal weight vector solution is obtained by MP criterion.

Because the actual covariance matrix is not ideal, the performance of the DMI algorithm is affected by the eigen-value spread of the covariance matrix. The divergence is determined by the temporal and spatial correlation between the desired signal and the interference or between the interference and interference.

The optimal weight vector by DMI algorithm can be computed as:

The K snapshots constitute data matrix $\boldsymbol{X}$, the covariance matrix $\boldsymbol{R}$ is given as

$$\boldsymbol{R}=\frac{\boldsymbol{X}\boldsymbol{X}^{\mathrm{H}}}{K} \tag{82}$$

Directly estimate the covariance matrix and then by matrix inversion, obtain the inverse matrix $\boldsymbol{R}^{-1}$ combined with the desired signal direction vector, and the optimal weight vector is calculated according to Eq. (69).

$$w=\frac{\boldsymbol{R}^{-1}\boldsymbol{a}_0}{\boldsymbol{a}_0^{\mathrm{H}}\boldsymbol{R}^{-1}\boldsymbol{a}_0} \tag{83}$$

DMI algorithm needs to choose suitable number of sampling snapshots K. When the number of snapshots K is sufficiently large, the covariance matrix $\boldsymbol{R}$ is more accurate, but larger number of sampling snapshots increases the computing load [34]. The major disadvantage of DMI algorithm is its computational complexity which makes it difficult to implement on FPGA and DSP. On the other hanf, the truncated finite number of computation makes the matrix inverse operation instable.

extremely simple and numerically robust.

6.2 Least mean square algorithm

The least mean square (LMS) algorithm proposed by Widrow et al. [20] is the most classical algorithm in signal processing. The LMS algorithm is extremely simple and numerically robust. More detailed description about the LMS algorithm is given in Ref. [18, 35]. The LMS algorithm is based on the method of steepest descent, and therefore sometime it is referred to as a Stochastic Gradient Descent (SGD) algorithm. The unconstrained LMS algorithm is a training sequence based adaptive spatial filtering algorithm which recursively compute and update the optimal weight vector. It uses the gradient search method to solve the weight vector, thus avoiding the direct matrix inversion of the covariance matrix. Its iterative equation is given as

$$\boldsymbol{w}(k+1) = \boldsymbol{w}(k) + \mu \boldsymbol{g}(\boldsymbol{w}(k)) \tag{84}$$

where $\boldsymbol{w}(k+1)$ represents the new weight vector computed at the $(k+1)^{th}$ iteration, $\boldsymbol{g}(\boldsymbol{w}(k))$ is the gradient vector of the squared error (objective function) with respect to the weight vector $\boldsymbol{w}(k)$, and the scalar constant μ is the step size parameter which controls the rate of convergence [33]. The gradient vector is given by

$$\boldsymbol{g}(\boldsymbol{w}(k)) = -2\boldsymbol{x}(k+1)\varepsilon^* (\boldsymbol{w}(k)) \tag{85}$$

where $\boldsymbol{x}(k+1)$ is the $k+1$ array snapshots, namely the $k+1$ array sample, and $\varepsilon^* (\boldsymbol{w}(k))$ is the error between the array output and the reference signal [33]. Thus, the estimated gradient vector is the product of the error between the array output and the reference signal, and the array signal received at the $k-th$ iteration. The error $\varepsilon^* (\boldsymbol{w}(k))$ can be expressed as

$$\varepsilon(\boldsymbol{x}(k)) = d(k+1) - \boldsymbol{w}^H(k)\boldsymbol{x}(k+1) \tag{86}$$

where $d(k+1)$ is the reference signal at the $(k+1)^{th}$ iteration. As one of the most classical adaptive filtering algorithms, ULMS has the advantage of computational simplicity and simple hardware requirement, but its convergence speed is relatively slow. In order to ensure the convergence of the algorithm, the iterative step size must meet the following condition [18, 20, 33–37].

$$0 < \mu < \frac{2}{\lambda_{\max}} \tag{87}$$

where $\lambda_{\max}$ denoted the largest eigenvalue of the received signal covariance matrix.

The algorithm is based on the gradient of the adaptive algorithm, which is an important feature of the gradient of the average value problem. The mean of the gradient estimate is expressed as

$$\overline{\boldsymbol{g}}(\boldsymbol{w}(k)) = 2\boldsymbol{R}\boldsymbol{w} - 2\boldsymbol{r}_{xd} \tag{88}$$

In the iterative process of the algorithm, the gradient vector can be obtained by estimation. From the mean or expected value of the gradient estimate, the estimate is unbiased. At the same time, the estimation of the variance has also an effect on the performance of the algorithm. The variance is defined as

$$\xi(\boldsymbol{w}(k)) = E\left\{ \left| d(k) - \boldsymbol{w}^H(k)\boldsymbol{x}(k+1) \right|^2 \right\} \tag{89}$$

whose value is the error between the reference signal and the array output signal. From this, we can see that the Misadjustment of LMS algorithm is

$$MA = \mu tr\left\{[\boldsymbol{I} - \mu\boldsymbol{R}]^{-1}\boldsymbol{R}\right\} \tag{90}$$

The misadjustment defined as a ratio provides a measure of how close an adaptive algorithm is to optimality in the mean-square-error sense. The smaller the misadjustment, the more accurate is the steady-state solution of the algorithm. In other words, the difference between the weights estimated by the adaptive algorithm and optimal weights is further characterized by the ratio of the average excess steady-state MSE and the MMSE. It is referred to as the misadjustment. It is a dimensionless parameter and measures the performance of the algorithm. The misadjustment is a kind of noise and is caused by the use of noisy estimate of the gradient [38, 39].

From the above analysis, we can see that the LMS algorithm has different performance when choosing different steps and different covariance matrix estimation methods.

The basic steps of the LMS algorithm are as follows:

1. First initialize, $\boldsymbol{w}(0) = 0$, $k = 0$;

2. Iterative updates, so that $k = k + 1$;

$$e(k+1) = d(k+1) - \mathbf{w}^{\mathrm{T}}(k)\mathbf{x}(k+1)$$
$$\mathbf{w}(k+1) = \mathbf{w}(k) + \mu\mathbf{x}(k+1)e(k+1)$$

3. Stop iteration after the weight vector $\boldsymbol{w}(k)$ is convergent, so this time define$k = K$,$\boldsymbol{w}(K)$ is the desired weight vector.

Figure 5 shows the learning curve of the LMS algorithm with different step size parameters. It can be seen that when the step size parameter μ is small, the algorithm converges slowly, while the large value of step size parameter μ make the algorithm converge faster.

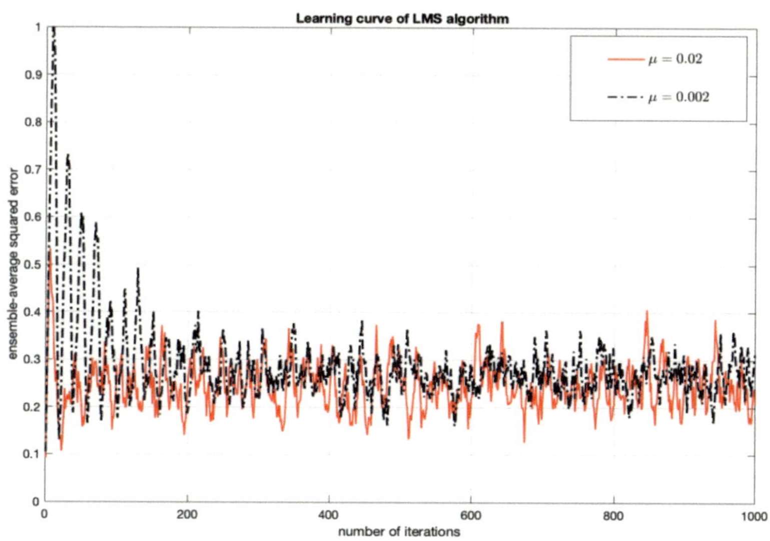

Figure 5.
Learning curve of the LMS algorithm.

The least mean square algorithm requires the training sequence, if the training sequence in the LMS algorithm is replaced by the DOA information of the desired signal, the Frost LMS algorithm can be obtained [40].

Iterative equation of the Frost LMS algorithm is

$$w(k+1) = P\{w(k) - \mu \mathbf{g}(w(k))\} + \frac{a_0}{L} \tag{91}$$

where the matrix

$$P = I - a_0 \left(a_0^{\mathrm{H}} a_0\right)^{-1} a_0^{\mathrm{H}} \tag{92}$$

and $g(w(k))$ is the gradient vector of the output signal power with respect to the weight vector $w(k)$, and is given by

$$g(w(k)) = x(k+1)y^*(k+1) \tag{93}$$

In the above equation, the output signal is given as

$$y(k+1) = w^{\mathrm{H}}(k)x(k+1) \tag{94}$$

Moreover, the initial value of the weights is given as

$$w(0) = \frac{a_0}{L} \tag{95}$$

In order to ensure the convergence of the iterative algorithm, the iterative step size still needs to meet the following conditions $\mu < 2/\lambda_{\max}$, where $\lambda_{\max}$ is the largest eigenvalue of the covariance matrix of the received signal.

Basic steps for the Frost LMS algorithm are as follows:

1. First initialize, $w(0) = \frac{a_0}{L}$, $k = 0$

2. Iterative updates, so that $k = k + 1$;

$$y(k+1) = \mathbf{w}^{\mathrm{H}}(k)\mathbf{x}(k+1);$$

$$\mathbf{w}(k+1) = \left(\mathbf{I} - \mathbf{a}_0\left(\mathbf{a}_0^{\mathrm{H}}\mathbf{a}_0\right)^{-1}\mathbf{a}_0^{\mathrm{H}}\right)\{\mathbf{w}(k) - \mu\mathbf{x}(k+1)y^*(k+1)\} + \frac{\mathbf{a}_0}{L};$$

3. Stop iteration after the weight vector $w(k)$ is convergent, so this time define $k = K$, $w(K)$is the desired weight vector.

The convergence rate of both the LMS algorithm and Frost LMS algorithm is associated with the step size parameter. Since, the eigenvalues of the received signal covariance matrix are not easy to obtain, the appropriate step size parameter cannot be chosen easily.

If the step size is too larger than twice the reciprocal of the maximum eigenvalue of the covariance matrix of the received signal, the weight vector diverges. Large μ's (step-size) speed up the convergence of the algorithm but also lower the precision of the steady-state solution of the algorithm. It should be noted that value of the step size must be less than twice the reciprocal of the maximum eigenvalue. Similarly, when the step-size is much less than twice the reciprocal of the maximum eigenvalue of the covariance matrix of received signals, the offset (steady state error) is small but the weight vector converges slowly.

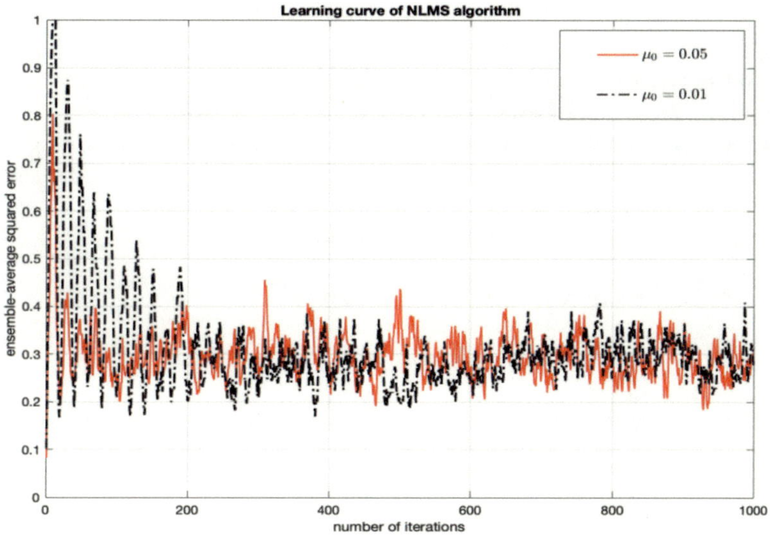

Figure 6.
Learning curve of the NLMS algorithm.

Another variant of the LMS family is the normalized LMS (NLMS) algorithm. This algorithm replaces the constant-step-size of conventional LMS algorithm with a data-dependent normalized step size at each iteration. At the k-th iteration, the step size is given by

$$\mu(k) = \frac{\mu_0}{x^H(k)x(k)} \tag{96}$$

where μ_0 is a constant. The .convergence of the NLMS algorithm is faster as compared to the LMS algorithm due to the data-dependent step size. **Figure 6** shows the convergence behavior of the NLMS algorithm with different μ_0.

One major advantage of the LMS algorithm is its simplicity, and when the step size is selected appropriately, the algorithm is stable (converged properly) and easy to be realized [21]. However, the LMS algorithm is sensitive to eigenvalues of the covariance matrix of received signals, and the convergence of the algorithm is poor when the eigenvalues are dispersed.

Various other variants of LMS algorithm are briefly discusses in [21]. In recent years, adaptive filtering algorithms have been extended into DOA estimation. DOA estimation based on adaptive filtering algorithms can be found in [41, 42].

6.3 Conjugate Gradient Method

The Conjugate Gradient Method (CGM) [43–45] proposed by Hestenes and Stiefel in 1952 (as direct method), is generally applied to the symmetric positive definite linear systems equations of the form $\boldsymbol{A}\boldsymbol{w} = \boldsymbol{b}$. In application of antenna arrays, the the weight vector computation by conjugate gradient method is discussed in [46]. Here, we have briefly outlined the conjugate gradient method (CGM) in application to beamforming [47].

In array signal processing, $\boldsymbol{w}$ represent the array weight vector, $\boldsymbol{A}$ is a matrix whose columns are corresponded to the consecutive samples obtained from array elements, while $\boldsymbol{b}$ is a vector containing consecutive samples of the desired signal. Thus, a residual vector

$$\boldsymbol{r} = \boldsymbol{b} - \boldsymbol{A}\boldsymbol{w} \tag{97}$$

refers to the error between the desired signal and array output at each sample, with the sum of the squared error given by $\boldsymbol{r}^H\boldsymbol{r}$.

The process is started with weight vector $\boldsymbol{w}(0)$ as an initial guess, to get a residual

$$\boldsymbol{r}(0) = \boldsymbol{b} - \boldsymbol{A}\boldsymbol{w}(0) \tag{98}$$

and the initial direction vector can be expressed as

$$\boldsymbol{g}(0) = \boldsymbol{A}^H\boldsymbol{r}(0) \tag{99}$$

Then moves the weights in this direction to yield a weight update equation

$$\boldsymbol{w}(k+1) = \boldsymbol{w}(k) + \mu(k)\boldsymbol{g}(k) \tag{100}$$

where the step size $\mu(k)$ is

$$\mu(k) = \frac{\left|\boldsymbol{A}^H\boldsymbol{r}(k)\right|^2}{\left|\boldsymbol{A}^H\boldsymbol{g}(k)\right|^2} \tag{101}$$

The residual $\boldsymbol{r}(k)$ and the direction vector $\boldsymbol{g}(k)$ are updated using

$$\boldsymbol{r}(k+1) = \boldsymbol{r}(k) + \mu(k)\boldsymbol{A}\boldsymbol{g}(k) \tag{102}$$

and

$$\boldsymbol{g}(k+1) = \boldsymbol{A}^H\boldsymbol{r}(k+1) - \alpha(k)\boldsymbol{g}(k) \tag{103}$$

with

$$\alpha(k) = \frac{\left|\boldsymbol{A}^H\boldsymbol{r}(k+1)\right|^2}{\left|\boldsymbol{A}^H\boldsymbol{r}(k)\right|^2} \tag{104}$$

A pre-determined threshold level is defined and the algorithm is stopped when the residual falls below the threshold level.

It should be noted that the direction vector points in the direction of error surface gradient $\boldsymbol{r}^H(k)\boldsymbol{r}(k)$ at the $k-th$ iteration, which the algorithm is trying to minimize. The method converges to the error surface minimum within at most K iterations for a K-rank matrix equation, and thus provides the fastest convergence of all iterative methods [46, 48].

6.4 Recursive least square algorithm

In order to further improve the convergence rate, a more sophisticated algorithm is recursive least square algorithm. RLS algorithm is based on the Recursive Least Squares Estimation (RLSE), which uses time average instead of statistical (ensemble) average or stochastic expectations. The RLS algorithm work well even when the eigenvalue spread of the input signal correlation matrix is large [49, 50]. So RLS algorithm has an advantage of insensitivity to variations in eigenvalue spread of the input correlation matrix [49, 50]. These algorithms have excellent

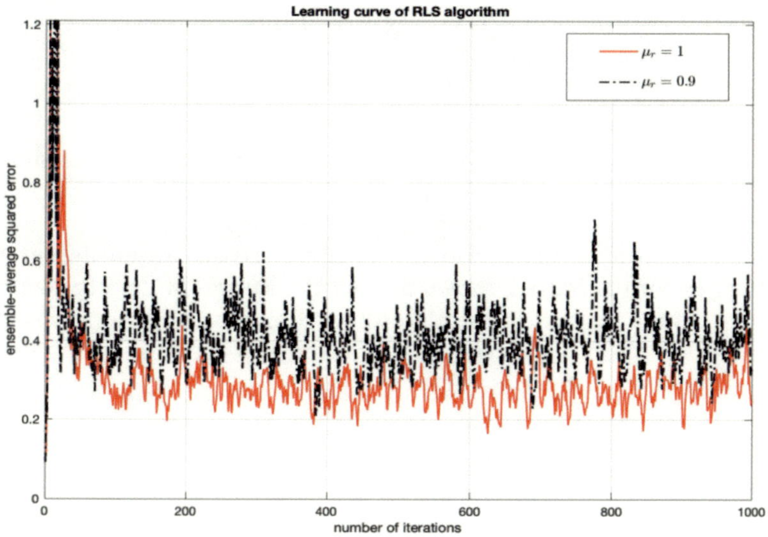

Figure 7.
Learning curve of the RLS algorithm.

performance when working in time-varying environments [49, 50]. Therefore, in the practical application, the forgetting factor μ is usually taken into account, and the optimal weight vector solution is slightly different. According to the optimal weight vector solution of MP criterion, the covariance matrix estimation is defined as

$$\boldsymbol{\Phi}(k) = \sum_{k=1}^{K} \mu^{K-k} \boldsymbol{x}(k)\boldsymbol{x}^H(k) \tag{105}$$

where the parameter μ should be chosen in the range $0 \ll \mu \leq 1$.

The above equation can also be expressed as

$$\boldsymbol{\Phi}(k) = \mu \boldsymbol{\Phi}(k-1) + \boldsymbol{x}(k)\boldsymbol{x}^H(k) \tag{106}$$

Using Matrix Inversion Lemma [14, 36, 51–54] (See Appendix A)

$$\begin{aligned} \boldsymbol{P}(k) &= \boldsymbol{\Phi}^{-1}(k) \\ &= \mu^{-1}\boldsymbol{\Phi}^{-1}(k-1) - \frac{\mu^{-2}\boldsymbol{\Phi}^{-1}(k-1)\boldsymbol{x}(k)\boldsymbol{x}^H(k)\boldsymbol{\Phi}^{-1}(k-1)}{1+\mu^{-1}\boldsymbol{x}^H(k)\boldsymbol{\Phi}^{-1}(k-1)\boldsymbol{x}(k)} \end{aligned} \tag{107}$$

Let

$$\mathbf{g}(k) = \frac{\mu^{-1}\boldsymbol{\Phi}^{-1}(k-1)\boldsymbol{x}(k)}{1+\mu^{-1}\boldsymbol{x}^H(k)\boldsymbol{\Phi}^{-1}(k-1)\boldsymbol{x}(k)} \tag{108}$$

then Eq. (106) can be expressed as

$$\boldsymbol{P}(k) = \mu^{-1}\boldsymbol{P}(k-1) - \mu^{-1}\mathbf{g}(k)\boldsymbol{x}^H(k)\boldsymbol{P}(k-1) \tag{109}$$

The iterative formula of the algorithm can be expressed as.

$$\begin{aligned} \boldsymbol{w}(k) &= \Lambda(k)\left[\mu^{-1}\boldsymbol{P}(k-1) - \mu^{-1}\boldsymbol{g}(k)\boldsymbol{x}^H(k)\boldsymbol{P}(k-1)\right]\boldsymbol{a}(\theta_0) \\ &= \left\{\frac{\Lambda(k)}{\mu\Lambda(k)-1}\left[\boldsymbol{I} - \boldsymbol{g}(k)\boldsymbol{x}^H(k)\right]\right\}\boldsymbol{w}(k-1) \end{aligned} \tag{110}$$

By taking different values of the K, the optimal weight vector recursion expression can be obtained. Compared with the LMS algorithm, RLS has a faster convergence rate, which is also a closed-loop adaptive algorithm.

The implementation of the RLS algorithm is carried out with different values of the forgetting factor μ. **Figure** 7 shows the learning curves of the RLS algorithm. With the forgetting factor $\mu = 1$, the algorithm requires only 50 iterations to converge to its steady-state. It takes only 25 adaptation cycles to converge the RLS algorithm with a lower forgetting factor of $\mu = 0.9$.

7. Conclusion

In this chapter, we have introduced the basic principles and theoretical background of narrowband array signal processing. In particular, this chapter emphasized the fundamentals of narrowband signal processing exclusively used for the narrowband beamforming and DOA estimation. Furthermore, we reviewed the geometry of adaptive array antennas, the mathematical approaches for the development of signal models of the receiver array, and the selection criteria of the received signal processing technique, i.e. the criteria and guidelines related to adaptive filtering algorithms for solving the optimal weights. Considering the far-field narrowband signal using a uniform linear array as an example, the mathematical model is established in this chapter for the adaptive array antenna beamforming system. The basic theory of this chapter also laid a foundation for the theory of the wideband signal beamforming, which is then convenient for us to understand.

Appendix A

Matrix Inversion Lemma [52]: Let $\boldsymbol{A}$ and $\boldsymbol{B}$ be two positive-definite $N \times N$ matrices, $\boldsymbol{C}$ a $N \times M$ matrix, and $\boldsymbol{D}$ a positive definite $M \times M$ matrix. If they are related by

$$\boldsymbol{A} = \boldsymbol{B} + \boldsymbol{C}\boldsymbol{D}^{-1}\boldsymbol{C}^T,$$

then the inverse of the matrix $\boldsymbol{A}$ is

$$\boldsymbol{A}^{-1} = \boldsymbol{B}^{-1} - \boldsymbol{B}^{-1}\boldsymbol{C}\left(\boldsymbol{D} + \boldsymbol{C}^T\boldsymbol{B}^{-1}\boldsymbol{C}\right)^{-1}\boldsymbol{C}^T\boldsymbol{B}^{-1}$$

Author details

Zeeshan Ahmad
School of Electronic and Information Engineering, Ningbo University of Technology, Ningbo, China

*Address all correspondence to: azee@nbut.edu.cn

References

[1] Haykin S (Ed.). Advances in spectrum analysis and array processing (Vol. III). Englewood Cliffs, New Jersey, USA: Prentice Hall, 1995.

[2] Pillai S U. Array signal processing. New York, USA: Springer-Verlag, 2002.

[3] Haykin S, Liu K J R. Handbook on array processing and sensor networks. NJ, Hoboken: Wiley-IEEE Press, 2010.

[4] Haykin S. Adaptive filter theory: International edition. 5th Ed., Pearson Education: Prentice Hall, 2014.

[5] Frank G. Smart antennas with MATLAB, 2nd ed.New York: McGraw-Hill Education ; 2015.

[6] Salous S. Radio Propagation measurement and channel modelling. Hoboken, NJ, USA: John Wiley & Sons, 2013.

[7] Shirvani-Moghaddam S., Akbari F. A novel ULA-based geometry for improving AOA estimation. EURASIP Journal on Advances in Signal Processing, 2011, 2011(1): 39.

[8] Haykin S, Justice J H. Array signal processing. Englewood Cliffs, New Jersey, USA: Prentice-Hall signal processing series, 1985.

[9] Han Y. A rao–blackwellized particle filter for adaptive beamforming with strong interference. IEEE Transactions on Signal Processing, 2012, 60(6): 2952-2961.

[10] Hou Y. Research on digital beamf orming algorithms [thesis]. Harbin: Harbin Institute of technology, 2010.

[11] Zhang X. Modern signal processing, 3rd ed. Beijing: Tsinghua University Press, 2015.

[12] Guo Y. Constant modulus algorithm and its application in blind beamforming [thesis]. Xi'an: Xidian University, 2001.

[13] Gooch R, Lundell J. The CM array: An adaptive beamformer for constant modulus signals. ICASSP ‘86. IEEE International Conference on Acoustics, Speech, and Signal Processing, 1986, 11: 2523-2526.

[14] Krim H, Viberg M. Two decades of array signal processing research: the parametric approach. IEEE Signal Processing Magazine, 1996, 13(4): 67-94.

[15] Rong Y, Eldar Y C, Gershman A B. Performance tradeoffs among beamforming approaches. 2006 IEEE Sensor Array and Multichannel Signal Processing Workshop Proceedings, SAM 2006., 2006, pp.26-30.

[16] Trees H, Di S, Tang J. Optimal array processing technology. Beijing: Tsinghua University Press, 2008.

[17] Compton R T. Adaptive Antennas: Concept and Performance. Englewood Cliffs, New Jersey, USA: Prentice Hall, 1988.

[18] Monzingo R A, Miller T W. Introduction to Adaptive Arrays. New York, USA: John Wiley & Sons, 1980.

[19] Bao C-H, Shui P-L. Fast System Identification Using Direct Matrix Inversion and a critically Sampled Subband Adaptive Filter. Journal of Electronics Information Technology, 2008, 30(1): 139-143.

[20] Widrow B, Stearns S D. Adaptive Signal Processing. Upper Saddle River, New Jersey, USA: Prentice Hall, 1985.

[21] Srar J A, Chung K S, Mansour A. Adaptive Array Beamforming Using a Combined LMS-LMS Algorithm. IEEE Transactions on Antennas and Propagation, 2010, 58(11): 3545-3557.

[22] Griffiths L J. A simple adaptive algorithm for real-time processing in antenna arrays. Proceedings of the IEEE, 1969, 57(10): 1696-1704.

[23] Bhotto MZA, Antoniou A. Robust Recursive Least-Squares Adaptive-Filtering Algorithm for Impulsive-Noise Environments. IEEE Signal Processing Letters, 2011, 18(3): 185-188.

[24] Eweda E, Macchi O. Convergence of the RLS and LMS adaptive filters. IEEE Transactions on Circuits and Systems, 1987, 34(7): 799-803.

[25] Eleftheriou E, Falconer D D. Tracking properties and steady-state performance of RLS adaptive filter algorithms. IEEE Transactions on Acoustics, Speech and Signal Processing, 1986, 34(5): 1097-1110.

[26] Ahmad N A. A Globally Convergent Stochastic Pairwise Conjugate Gradient-Based Algorithm for Adaptive Filtering. IEEE Signal Processing Letters, 2008, 15: 914-917.

[27] Choi S, Kim D H. Adaptive antenna array utilizing the conjugate gradient method for compensation of multipath fading in a land mobile communication. [1992 Proceedings] Vehicular Technology Society 42nd VTS Conference - Frontiers of Technology. , 1992, 1: 33-36.

[28] Treichler J, Agee B. A new approach to multipath correction of constant modulus signals. IEEE Transactions on Acoustics, Speech, and Signal Processing, 1983, 31(2): 459-472.

[29] Elnashar A, Elnoubi S, Elmikati H. Sample-by-sample and block-adaptive robust constant modulus-based algorithms. IET Signal Processing, 2012, 6(8): 805-813.

[30] Shynk J J, Chan C K. Performance surfaces of the constant modulus algorithm based on a conditional Gaussian model. IEEE Transactions on Signal Processing, 1993, 41(5): 1965-1969.

[31] Rebeiz E, Urriza P, Cabric D. Optimizing Wideband Cyclostationary Spectrum Sensing Under Receiver Impairments. IEEE Transactions on Signal Processing, 2013, 61(15): 3931-3943.

[32] Xi J, Han W. Application of high-order cumulant in the phase-space reconstruction of multivariate chaotic series. Proceedings of 2010 International Conference on Intelligent Control and Information Processing, ICICIP 2010., Dalian, 2010, pp. 49-53.

[33] Hanzo L, Blogh J, Ni S. 3G, HSPA and FDD versus TDD Networking: Smart Antennas and Adaptive Modulation. 2nd ed., Wiley-IEEE Press, 2008.

[34] Zaharov VV, Teixeira M. SMI-MVDR Beamformer Implementations for Large Antenna Array and Small Sample Size. IEEE Transactions on Circuits and Systems I: Regular Papers, 2008, 55(10): 3317-3327.

[35] Widrow B, Mantey P E, Griffiths LJ, Goode BB. Adaptive antenna systems. Proceedings of the IEEE, 1967, 55(12): 2143-2159.

[36] Hollemans W, Prasad R, Kegel A. Performance analysis of cellular digital mobile radio systems including diversity techniques. The 8th IEEE International Symposium on Personal, Indoor and Mobile Radio Communications. Waves of the Year 2000. PIMRC '97., 1997, 2: 266-270.

[37] Winters J H, Salz J, Gitlin R D. The impact of antenna diversity on the capacity of wireless communication systems. IEEE Transactions on Communications, 1994, 42(234): 1740-1751.

[38] Widrow B. Adaptive filters I: Fundamentals. Technical Report 6784–

8, Stanford Electrical Engineering Laboratory, Stanford University, Stanford, CA, December 1966.

[39] Godara L C. Smart Antennas (Electrical Engineering & Applied Signal Processing Series). 1st ed., Boca Raton, Fla: CRC Press, 2004.

[40] Ahmad, Z., Yaoliang, S., & Du, Q. Adaptive wideband beamforming based on digital delay filter. Journal of Microwaves, Optoelectronics and Electromagnetic Applications, 2016, 15 (3), 261-274.

[41] Zeng, H., Ahmad, Z., Zhou, J., Wang, Q., Wang, Y. DOA estimation algorithm based on adaptive filtering in spatial domain. China Communications, 2016, 13(12): 49-58.

[42] Liu, C., Zhao, H. Efficient DOA Estimation Method Using Bias-compensated Adaptive Filtering. IEEE Transactions on Vehicular Technology, 2020. doi: 10.1109/TVT.2020.3020946.

[43] Hestenes M R, Stiefel E. Method of conjugate gradients for solving linear systems. Journal of Research of the National Bureau of Standards, 1952, 49 (6): 409–436.

[44] Daniel J W. The Conjugate Gradient Method for Linear and Nonlinear Operator Equations. SIAM Journal on Numerical Analysis, 1967, 4(1): 10–26.

[45] Sarkar T, Siarkiewicz K, Stratton R. Survey of numerical methods for solution of large systems of linear equations for electromagnetic field problems. IEEE Transactions on Antennas and Propagation, 1981, 29(6): 847-856.

[46] Choi S, Sarkar T K. Adaptive antenna array utilizing the conjugate gradient method for multipath mobile communication. Signal Processing, 1992, 29(3): 319-333.

[47] Godara L C. Application of antenna arrays to mobile communications. II. Beam-forming and direction-of-arrival considerations. Proceedings of the IEEE, 1997, 85(8): 1195-1245.

[48] Choi S. Application of the Conjugate Gradient Method for Optimum Array Processing. Progress in Electromagnetics Research, 1991, 5: 589–624.

[49] Diniz P S R. Adaptive Filtering: Algorithms and Practical Implementation. New York, USA: Springer US, 2012, pp.209-247

[50] Petroit R G, Ucci D R. A LMS and RLS look-ahead adaptive array for spread spectrum signals. Antennas and Propagation Society International Symposium. AP-S. Merging Technologies for the 90's. Digest., Dallas, TX, USA, 1990, 4: 1442-1445.

[51] Ceng W H W T. Cordless Telecommunications in Europe. London, U.K: Springer-Verlag London, 1990.

[52] Ochsner H. The digital european cordless telecommunications specification, DECT. in Tuttlebee W H W (Ed.) . Cordless telecommunications in Europe: the evolution of personal communications., Springer-Verlag New York, NJ, 1991, pp. 273–285.

[53] Petrus P, Reed J H, Rappaport T S. Effects of directional antennas at the base station on the Doppler spectrum. IEEE Communications Letters, 1997, 1 (2): 40-42.

[54] T. Kailath, Linear Systems. Englewood Cliffs, New Jersey, USA: Prentice-Hall, 1980.

Chapter 2

Reconfigurable Filter Design

Tae-Hak Lee, Sang-Gyu Lee, Jean-Jacques Laurin and Ke Wu

Abstract

This chapter discusses recent development of reconfigurable filters. The technical terminology reconfigurable means that a circuit is designed in a way to have various electrical characteristics comparing with one which has a static feature. For the filter design, the various electrical characteristics can be considered as the filter can tune its operating frequency, bandwidth, and/or have multiple operational modes, that is, bandstop or bandpass modes. Also, recently, the filters that can exhibit an improved impedance matching performance over its stopband have been reported. It provides more options for the filter designers to realize the reconfigurable filters having reflective and/or absorptive frequency response types to satisfy a prior given requirement. In this chapter, recently devised filter designs will be covered and essential frequency tuning elements to realize the reconfigurable characteristic will be introduced as well.

Keywords: resonant frequency, operational modes, reflective, absorptive, frequency tuning elements

1. Introduction

Microwave filters play an important role in the chain of radio frequency (RF) front end to transmit and receive the required signals or to block the undesired ones. Most filter designs are dependent on the electrical length of the operating frequency or the field configuration of the resonant modes inside a cavity so the reconfigurable characteristic such as the capability to tune the operating frequency, bandwidth, and operational modes can be obtained by controlling the dependent design parameters. Recently, some researchers embarked on the development of filters that have the better matching performance at its stopband region to avoid using the isolators. The improved impedance matching characteristic is achieved in a way the input signal not to be reflected back to the input port by absorbing the input signal inside the filter structure so the devised circuits having the improved impedance performance are often named as an absorptive or a reflectionless filter.

In this chapter, we will explore the recent development of reconfigurable filter designs that can change their operating frequency, bandwidth, and operational modes along with some tuning components. Besides, the filter designed to have both reflective and absorptive characteristics will be shown. The frequency tunable substrate integrated waveguide (SIW) resonators are used to change the operating frequency of the reconfigurable filter and, to achieve frequency agility, the tuning elements based on the piezoelectric disk or the electromagnet are also given. To verify the tuning method using the electromagnet, frequency tunable filtering balun is fabricated and tested using four electromagnets. In the following section, we first start from the theoretical modelings of the frequency responses of the reconfigurable filter using coupling coefficients and equivalent circuits, and then the circuit model with simulation and measurement results are given to support the theoretical modelings.

2. Filter designs

The coupling coefficients and their array in form of a matrix containing the inverter values are widely used to explain or represent the operation mechanism of the filter structure. The fundamental theory and detailed concepts to establish the coupling coefficients and matrix can be found from well-known textbooks [1, 2]. In this section, we will briefly cover the definition of two coupling coefficients and the experimental process to assess the coupling coefficients of the physical external, or inter-resonator coupling structures. Both couplings essentially need to be defined for the theoretical frequency responses and the associated coefficient to each coupling structure should be realized from the coupling structures for the filter to satisfy the requirements.

2.1 Coupling coefficients modeling

In this subsection, two kinds of the coupling coefficients, external and internal, are explained and the way to obtain two different coefficients from the simulation or measurement process with ease will be analytically given.

Figure 1 shows an nth order filter circuit with serially connected LC resonators. The couplings between the resonators and between a resonator and input/output ports are shown with K-inverters and the values for the inverters are given using $K_{i,j}$ and $K_{0,1}$, respectively. Note that the given nth order circuit consists of the serially connected resonators and impedance inverters but the identical nth order frequency response can also be realized using parallel connected LC resonators and admittance inverters. In other words, in this chapter, we extract the theoretical frequency responses using the series LC resonators with K-inverters but the same results can be obtained with parallel resonators with J-inverters due to the duality theorem.

The inverter values for the required bandpass responses can be defined as follows [3],

$$K_{0,1} = \sqrt{\frac{Z_0 L_1}{g_0 g_1}}\Delta\omega, \quad K_{i,j} = \sqrt{\frac{L_i L_j}{g_i g_j}}\Delta\omega, \quad K_{n,n+1} = \sqrt{\frac{Z_0 L_n}{g_n g_{n+1}}}\Delta\omega \tag{1}$$

where the $\Delta\omega$ represents the bandwidth of the filter and the lowpass prototype elements are given using g_n. When the series LC resonators are replaced with generalized resonators which have the reactance slope parameter, x_i, and the inverter values given in Eq. (1) can be driven as Eq. (2),

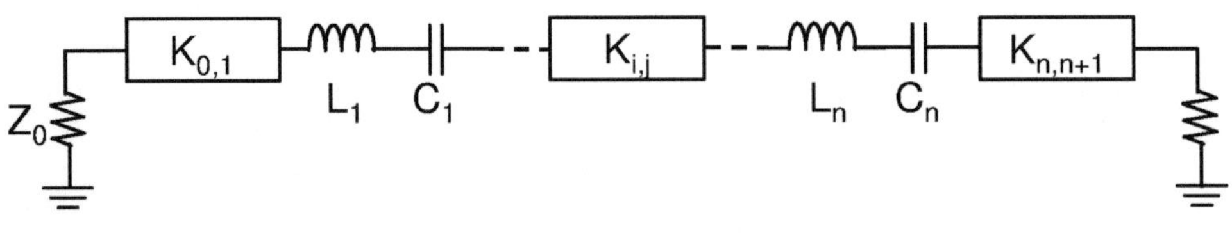

Figure 1.
Nth order filter structure composed of series LC components.

$$K_{0,1} = \sqrt{\frac{Z_0 x_1}{g_0 g_1}} \text{FBW}, \quad K_{i,j} = \sqrt{\frac{x_i x_j}{g_i g_j}} \text{FBW}, \quad K_{n,n+1} = \sqrt{\frac{Z_0 x_n}{g_n g_{n+1}}} \text{FBW}. \tag{2}$$

where the FBW stands for the fractional bandwidth. In Eq. (2), port impedance Z_0 can be normalized and the lowpass prototype elements, g-parameters, can be replaced using the normalized coupling coefficients, $M_{n,n+1}$. As a result, the K-inverter values of the nth order bandpass filter for the external and inter-resonator coupling structures are given in Eq. (3).

$$K_{0,1} = \sqrt{x_1 \text{FBW}} M_{0,1}, \quad K_{i,j} = \sqrt{x_i x_j} \text{FBW} M_{i,j}, \quad K_{n,n+1} = \sqrt{x_n \text{FBW}} M_{n,n+1}. \tag{3}$$

In a similar way, J-inverter values can be obtained with the susceptance slope parameters of parallel resonators and those are given in Eq. (4).

$$J_{0,1} = \sqrt{b_1 \text{FBW}} M_{0,1}, \quad J_{i,j} = \sqrt{b_i b_j} \text{FBW} M_{i,j}, \quad J_{n,n+1} = \sqrt{b_n \text{FBW}} M_{n,n+1}. \tag{4}$$

The theoretical frequency responses based on the normalized coupling coefficients, $M_{n,n+1}$, can be generated from the lowpass prototype g-parameters, g_n. **Figure 2** shows the theoretical frequency responses of the second-order filtering structures. The coupling coefficient matrices associated with the structure are given in the inset of the figures. Note that the filter can produce either bandpass or bandstop frequency response according to the coupling scheme. The theoretical responses are generated with an assumption that the filters are designed and realized with resonators having a quality factor of 350. The fractional bandwidth of the filter is set as 0.023. Note that both frequency responses are designed to have the 3-dB bandwidth at the normalized frequency of ω is equal to 1.

In order to meet the requirement regarding the frequency response of the reconfigurable filter, the coupling structure should be designed based on the simulation or measurement process to find a suitable value for each coupling structure. With the given values for the normalized coupling coefficients and the fractional bandwidth, the inverter values for both external and internal coupling structures need to be determined to realize the required filtering responses.

Figure 3 shows a simulation or measurement setup for the external coupling and its reflection coefficient result. The input impedance seen from the source and the reflection coefficient, Γ, can be given in Eq. (5).

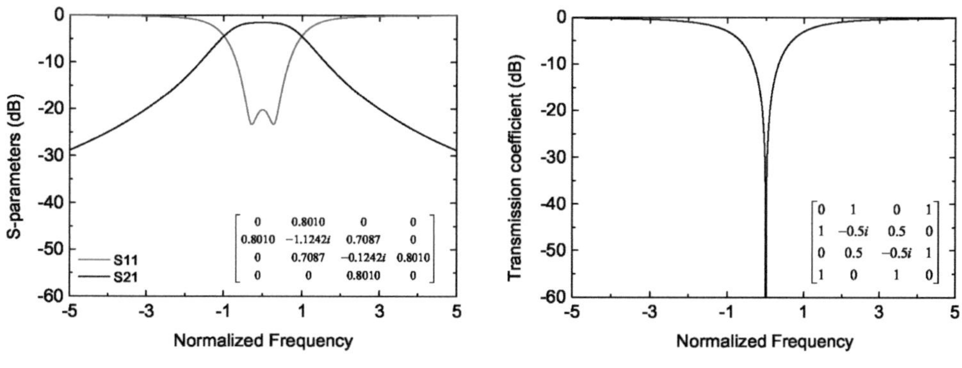

Figure 2.
Theoretical frequency responses.

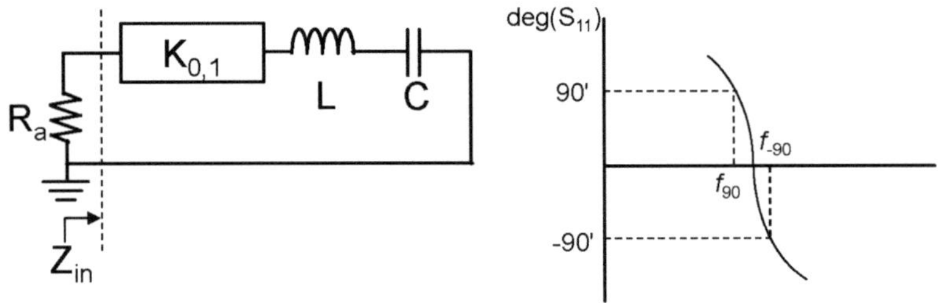

Figure 3.
Simulation setup for an external coupling coefficient and its phase response of the reflection coefficient.

$$Z_{in} = \frac{K_{0,1}^2}{j\omega_0 L\left(\frac{\omega}{\omega_0} - \frac{\omega_0}{\omega}\right)}, \qquad \Gamma = \frac{Z_{in} - R_a}{Z_{in} + R_a}. \tag{5}$$

The reflection coefficient can be reorganized with the input impedance, Z_{in}, and the frequency points at which the phase of the reflection coefficient meets $\pm 90°$. In other words, the reflection coefficient can be organized with respect to ω when its phase is $90°$ or $-90°$, and two positive solutions are given in Eq. (6).

$$\omega_{+90} = \frac{-k_{01}^2 + \sqrt{k_{01}^4 + 4(R_a w_0 L)^2}}{2R_a L}, \qquad \omega_{-90} = \frac{k_{01}^2 + \sqrt{k_{01}^4 + 4(R_a w_0 L)^2}}{2R_a L}. \tag{6}$$

So, the difference between two frequency points can represent the inverter value and, with the predefined form of Eq. (3), it can be summarized as Eq. (7) when the source impedance, R_a, is the same as the port impedance, Z_0.

$$f_{-90} - f_{+90} = \Delta f \cdot M_{0,1}^2. \tag{7}$$

where Δf is the bandwidth in Hz.

It means that the physical external coupling structure can be tuned during simulation or measurement process to meet the required design value. The design goal can be calculated with the given values such as the normalized coupling coefficient and bandwidth. As a result, the dimensions for external coupling structure can be determined or fine-tuned to achieve the prescribed frequency responses.

In addition to the external coupling structure, the reconfigurable filter also possesses the internal coupling structures and the inter-resonator coupling coefficients represent its coupling strength between resonators. **Figure 4** shows the equivalent circuit for the simulation or measurement setup for the inter-resonator coupling and its transmission coefficient result. As shown in the figure, two resonators are coupled each other through a coupling structure modeled with an inverter whose value is $K_{i,j}$, and both resonators are fed from source or load with loosely coupled through the inverters, K' or K''. In other words, to minimize any likely effects from input and output ports given with Z_0 in **Figure 4** on the inter-resonator coupling coefficients, the simulation or measurement setup for the inter-resonator coupling coefficients is designed to have small K' or K'' values. The input impedance seen from the loosely coupled external ports is given in Eq. (8).

$$Z_{in} = j\omega_0 L_i\left(\frac{\omega}{\omega_0} - \frac{\omega_0}{\omega}\right) + \frac{K_{i,j}^2}{j\omega_0 L_j\left(\frac{\omega}{\omega_0} - \frac{\omega_0}{\omega}\right)}. \tag{8}$$

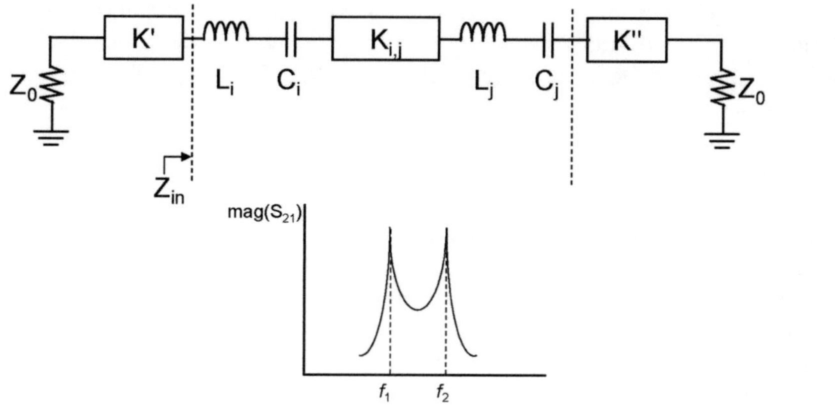

Figure 4.
Simulation setup for an inter-resonator coupling coefficient and its magnitude response of the transmission coefficient.

Since two resonant peaks in the transmission response given in **Figure 4** coincide with the short-circuited frequencies, we can have frequency points by calculating its zeros of Eq. (8) with respect to the ω. Each positive solution from two different equations can be given as f_1 and f_2.

$$f_1 = \frac{-K_{i,j} + \sqrt{K_{i,j}^2 + 4w_0^2 L_i L_j}}{4\pi\sqrt{L_i L_j}}, \quad f_2 = \frac{K_{i,j} + \sqrt{K_{i,j}^2 + 4w_0^2 L_i L_j}}{4\pi\sqrt{L_i L_j}}. \tag{9}$$

Similar with the case for the external coupling coefficient design, the inter-resonator coupling coefficient can also be estimated from the distance between two frequency points and it can be calculated as Eq. (10) with the theoretical values of $K_{i,j}$ with ease.

$$f_2 - f_1 = \Delta f \cdot M_{1,2}. \tag{10}$$

Based on Eqs. (7) and (10), one can estimate the coupling structures for both external and internal couplings and precisely optimize the dimensions of structures to realize the required filter responses.

Up to this, we designed coupling structures for the external and inter-resonator couplings, and it could be done by using both theoretical responses and simulation or measurement processes. The reconfigurable characteristic can be obtained by applying the electronic components such as the varactor or pin diodes to the static coupling structures. For example, the shunt-connected varactor diodes can be embedded in the inter-resonator coupling structure and it results in different coupling coefficients comparing with the one without varactor diodes. The external coupling coefficients can also be tuned with electronic components loading to provide the proper impedance matching performance. In the following subsection, the equivalent circuits of coupling structures with tuning elements are presented in more detail.

2.2 Equivalent circuit modeling

In the previous subsection, the theoretical coupling values comprising a filter structure are given and simulation or measurement setup to extract the coupling coefficients is also provided. In order to realize the required filter responses with the fabricated filter, we establish the equivalent circuits based on the theoretical values and also perform the full-wave simulation process with commercial tools such as Keysight Advanced Design System and Ansys High Frequency Structure Simulator. In this subsection, two equivalent circuits for the different frequency responses, frequency tunable bandpass and absorptive bandstop, are given to describe the operation mechanism of a reconfigurable filter [4]. Both equivalent circuits consist of the coupling structures that are covered in the previous section and those contain the electronic components to achieve the required reconfigurable characteristic. During the simulation, the capability of loaded electronic components that change the coupling coefficients can be figured out and the detailed simulation results are given in [4].

Figure 5 shows an equivalent circuit for a frequency tunable two-pole bandpass response. It contains two LC resonators coupled each other through an iris which is modeled using an inductor, L_C. A short-circuited microstrip line with shunt connected capacitors also contributes the inter-resonator coupling at the same time. The input and ouput lines are also coupled to the resonators. As mentioned in the previous section, the reconfigurable characteristics can be realized by tuning the coupling structures and, in this equivalent circuit, the short-circuited capacitors, C_i and C_e, are placed to optimize the internal and external coupling coefficients, respectively. The operating frequency agility is basically achieved by changing the value of C_r and the passband bandwidth can mainly be tuned by C_i. A capacitor serially shunt-connected to the input and output port using a short length of microstrip line (l_m) can control the matching performance. As a result, all features for the bandpass frequency responses can be controlled with three different capacitors.

Figure 6 shows the simulation results for the operating and bandwidth tuning performance of the bandpass equivalent circuit. The detailed circuit parameters can

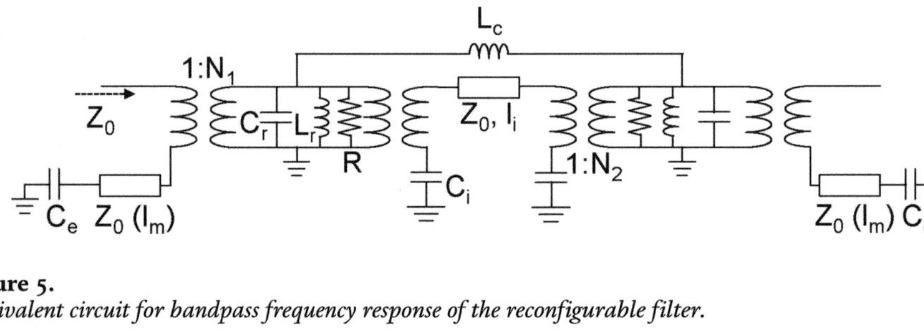

Figure 5.
Equivalent circuit for bandpass frequency response of the reconfigurable filter.

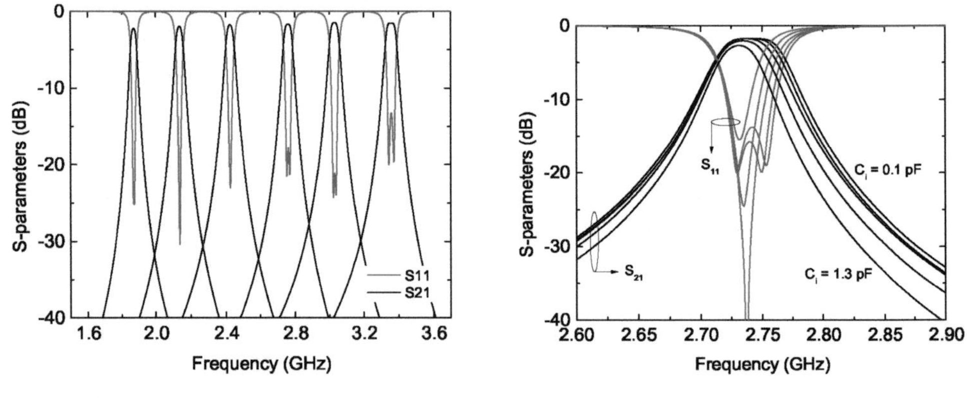

Figure 6.
Simulation results for the operating frequency and bandwidth tuning performance of the bandpass equivalent circuit.

be initially decided when the coupling structures for external and inter-resonator couplings are investigated using the full-wave simulation process as given in the previous section. The initial circuit parameters except for the three capacitance values are N_1 = 1.05, N_2 = 2.2, l_m = 0.357 in, l_i = 0.315 in, L_r = 10.467 nH and L_c = 6.978 nH. The capacitance for resonators C_r is used from 22 pF to 70 pF and the other two capacitance values are determined based on the reasonable values from the commercial varactor diodes. The frequency tuning ratio of larger than 1.7:1 is expected based on the simulation results. The different capacitance loading C_i changes the resonant frequency of a resonant mode so it results in the bandwidth control capability. Note that the simulation for the bandwidth tuning is performed with a fixed C_r value and the similar tuning performance can be obtained over the frequency tuning range of interest. The simulation results exhibit that one can tune the center frequency and bandwidth of the filter maintaining the impedance matching performance.

Figure 7 shows the equivalent circuit for the absorptive bandstop frequency response. Similar to the one for the bandpass frequency response it can change the operating frequency by changing the capacitance which is the part of resonators. In order to realize the unity coupling value between source and load as given in the inset of the theoretical frequency response, the microstrip line which is designed to have an electrical length of 270° at 2.2 GHz is added between two ports. Except for the microstrip line for M_{sl}, the rest of the circuit parameters are the same as those for the bandpass mode equivalent circuit. The absorptive frequency responses given in **Figure 8** can be characterized with the reduced reflection coefficients at the stopband of the bandstop filtering response. The equivalent circuit shows the best absorptive characteristic at 2.2 GHz of the predetermined center frequency but the amount of the reflection coefficient increases as the operating frequency is tuned from the predetermined one since the electrical length of the microstrip line deviates from the center frequency. In other words, the frequency tuning range of the absorptive bandstop mode is mainly limited by the electrical length of the filter structures and, in this case, it is about 500 MHz with less than −10 dB reflection coefficient.

Two equivalent circuits shown in **Figures 5** and **7** are realized using frequency tunable substrate integrated waveguide (SIW) resonators, varactor diodes, and single pole double throw (SPDT) switches. Two frequency tunable resonators and the microstrip line structure are placed at the different substrate and are coupled through coupling slots. The commercial varactor diodes are embedded in the microstrip line and are used to tune the coupling strength with different input voltages. In order to realize both bandpass and absorptive bandstop responses using a filter structure, two SPDT switches are also embedded in the microstrip line designed to provide M_{sl} coupling between source and load.

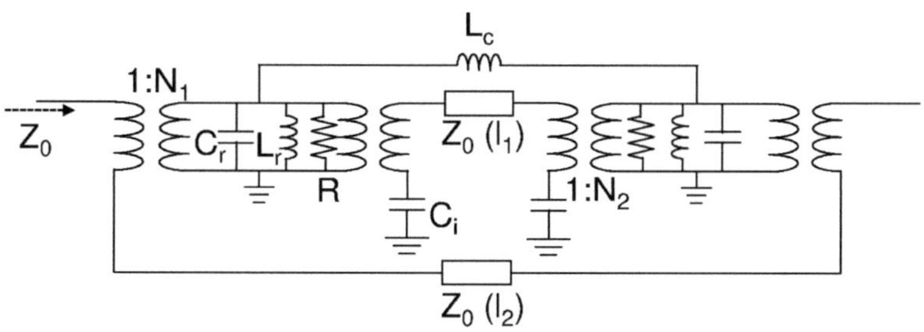

Figure 7.
Equivalent circuit for absorptive bandstop frequency response of the reconfigurable filter.

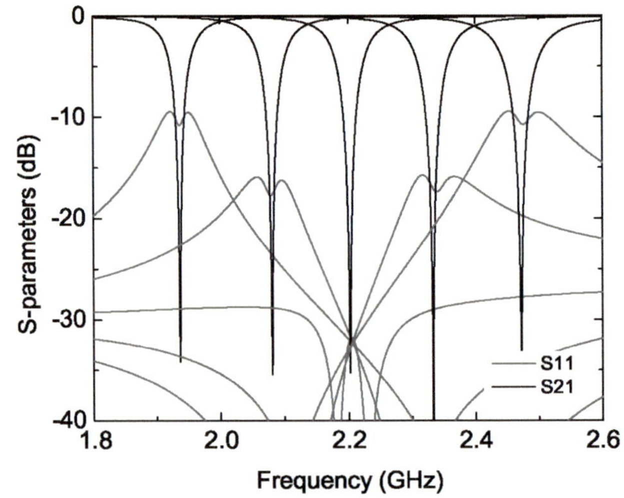

Figure 8.
Simulation results of the equivalent circuit for the absorptive bandstop responses.

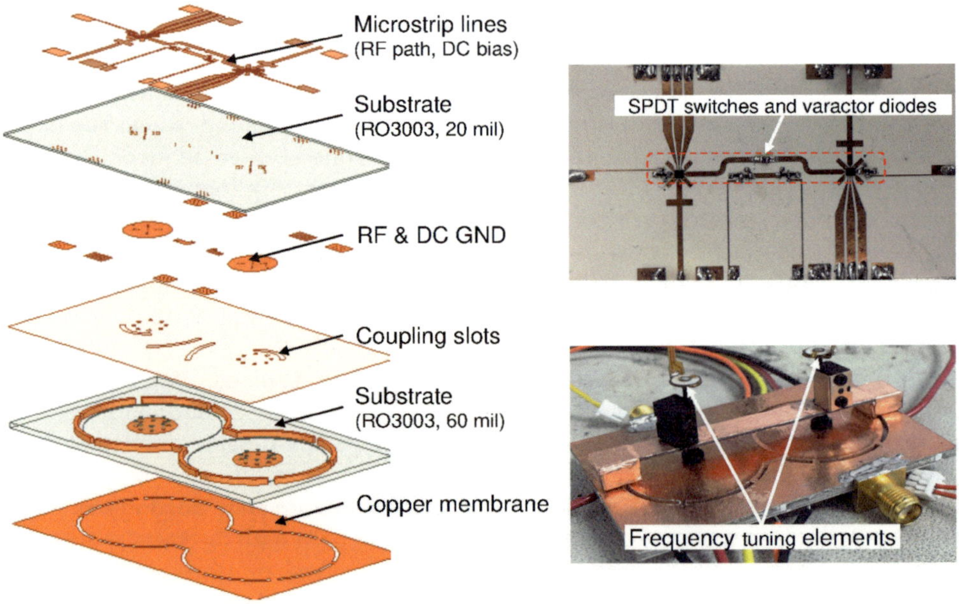

Figure 9.
Reconfigurable filter configuration and a photograph of the filter with frequency tuning elements.

2.3 SIW resonator-based reconfigurable filter

Figure 9 shows the filter configuration and a photograph of the fabricated filter with two frequency tuning elements [4]. At the top of the fabricated circuit, the microstrip line for the RF input signal is etched together with direct current (DC) bias lines for the electronic components. The coupling slots in the ground plane of the microstrip line are used to control the external or internal coupling and their size should be optimized in order to satisfy both bandpass and absorptive bandstop frequency responses. In virtue of the shunt-connected varactor diodes embedded in the microstrip, one can realize or achieve the coupling coefficient requirement over the frequency tuning range of interest. The SIW resonator consists of conductive posts at the center of the resonator and, with a copper membrane attached at the bottom of the resonator, high capacitance is loaded between the copper membrane and a circular-shaped ground plane of the post. The frequency agility is achieved by changing the loaded capacitance, and it means that the thickness of the air gap determines the resonant frequency of the frequency tunable SIW resonator. The frequency tuning elements attached at the bottom of the resonator are designed to change the air-gap thickness with the movement of a shaft connected to a piezoelectric disk actuator. More about the tuning element will be given in the next section.

Figure 10 shows the measurement results of the bandpass mode of the reconfigurable filter. The operating frequency is tuned from 1.86 to 3.3 GHz. The resonant frequency is tuned using two piezo disk-based tuning elements and the impedance matching performance of at least below −10 dB reflection coefficient is maintained over the frequency tuning range thanks to the varactor diodes for the external coupling optimization. As shown in the frequency tuning measurement results, the filter exhibits the different passband bandwidth as its operating frequency is tuned. Since the coupling strength generated from the coupling slots is unavoidably frequency dependent, so the coupling coefficient from a coupling slot with a fixed dimension can be different as the operating frequency of the filter is tuned. As mentioned earlier, the passband bandwidth is mainly controlled with the varactor diodes which have a capacitance value of C_i in **Figure 5**. The fabricated filter can also maintain its passband bandwidth even though the operating frequency is changed. In the right graph of **Figure 10**, the capacitance values are set to realize 80 MHz constant bandwidth with about 640 MHz frequency tuning range. **Figure 11** shows the measurement results of the absorptive bandstop mode of the fabricated filter. As expected from the simulation result given in **Figure 8**, the absorptive characteristic is maintained over the frequency tuning range of interest and the amount of reflection coefficient and attenuation performance get worse when the operation frequency moves away from the frequency where the l_2 given in **Figure 7** satisfies the electrical length of 270°. The fabricated filter provides more than 500 MHz of frequency tuning range when it operates as an absorptive bandstop mode.

In the following section, the frequency tuning elements that can be applied to the aforementioned frequency tunable SIW resonators will be given. First, the piezoelectric disk-based tiny ultra-linear actuator will be introduced with its detailed operating mechanism and then we will provide the recently proposed tuning method based on the magnetically actuated tuning elements. Finally, a filtering balun structure is fabricated and its frequency tunable characteristic is tested using four electromagnets to verify the proposed magnetically actuated tuning method.

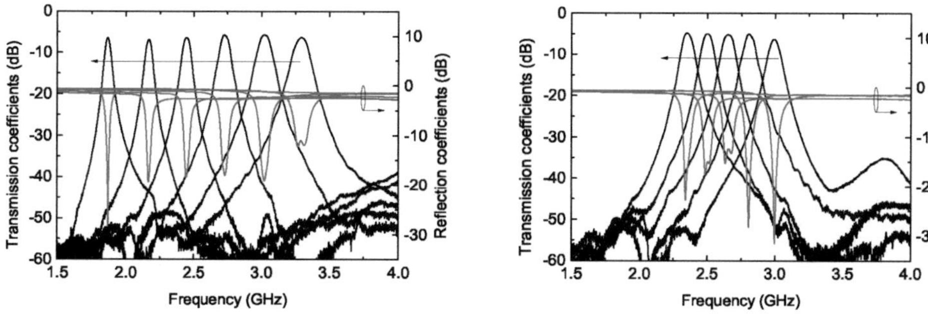

Figure 10.
Measured bandpass mode frequency responses.

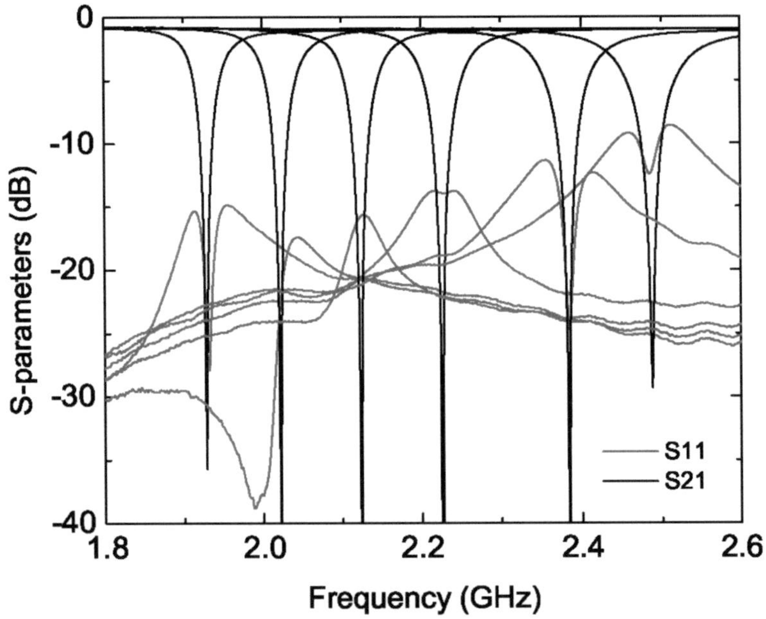

Figure 11.
Measured absorptive bandstop mode frequency responses.

3. Frequency tuning elements

In order to achieve frequency agility, one can exploit the tuning elements in the filter structures. Those can be varactor or pin diodes when the filters are designed with lumped or distributed elements such as microstrip lines since they can change and perturb the electrical length or the electromagnetic fields in the structures [5]. In this section, two different kinds of frequency tuning elements and their application to the frequency tunable substrate integrated waveguide resonator will be covered.

3.1 Piezoelectric disk-based elements

The frequency tunable SIW resonator shown in the previous section utilizes a piezoelectric disk-based actuator to change the thickness of an air gap in which the electric field is strong [6]. The piezoelectric actuator can be applied to the SIW resonator in two different ways.

Firstly, as shown in [6], the piezo disks can be directly attached to the copper membrane and react to the applied DC voltage. The thickness of the disk can be varied with the applied voltage so the different air-gap thicknesses result in the resonant frequency tuning. However, there are two drawbacks in directly attaching the piezo disk to the resonators. One is the size of the disk itself as it should cover the whole of the copper membrane and supporting substrate to properly operate its function and it can limit the size of the resonator as well. It means that the piezo disk may not be large enough to cover the frequency tunable resonator designed to operate at lower frequency bands. The second drawback is the large input voltage range with the hysteresis effects. The applied input voltage is dependent on the piezo disks, but it could be from −200 to +200 V to satisfy the sufficient frequency tuning range requirements. In addition, the amount of changes in the thickness is not identical when the applied voltage moves from low to high level or from low to high level due to the hysteresis effect.

Secondly, the frequency tunable substrate integrated waveguide resonator utilizes the tiny ultra-linear actuators, named TULA to tune the resonant frequency [7]. The devised element shown in **Figure 12** is comprised of a small piezo disk with a post attached to and the input voltage is applied by using a small driver circuit that can be controlled with commercial software. The applied voltage pulse has a form of sawtooth and the post attached to the piezo disk moves upward or downward. The reported linear actuator has an advantage over the disk type of piezoelectric actuator as it can provide a predictable amount of movement per the amplitude of pulse in spite of the hysteresis of the piezo disk. But these tuning elements can limit the filter to be assembled with a neighboring circuit due to the size of the fixture with a shaft. In addition, one end of the post should be glued to the copper membrane to control the air-gap thickness, so during the fabrication process, especially in the attachment of the post to the filter, there is a chance of serious damage to the copper membrane. In addition, it is not a practical way to realize the frequency tunable characteristic when low cost, compact volume designs are needed.

3.2 Magnetically actuated tuning elements

In this subsection, a recently reported frequency tuning method that can be applied to the frequency tunable substrate integrated waveguide resonators will be covered [8]. An electromagnet with a high permeability foil are utilized to tune the resonant frequency without any contacts between the resonators and tuning elements. A foil sheet is glued on the copper membrane during the fabrication process so the thickness of an air gap can be tuned with the applied magnetic flux from the electromagnet. Based on this method, the resonant frequency of the filter can be precisely tuned and the copper membrane can maintain its status as it was fabricated since the frequency tuning element, electromagnet, does not contact the filter circuit, unlike the piezoelectric actuator. A detailed explanation along with the simulation and measurement results is given in the following.

Figure 13 shows simplified 3D and side view of the circular-shaped substrate integrated waveguide resonator. A copper plate is circularly etched at the bottom of the substrate and is also electrically connected to the top plate via conductive via-holes. A large amount of capacitance is generated at the air gap between the copper

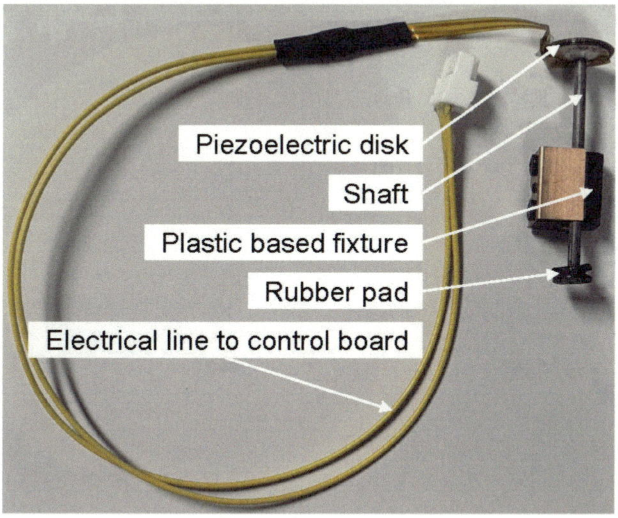

Figure 12.
A photograph of a piezoelectric disk-based tiny ultra-linear actuator.

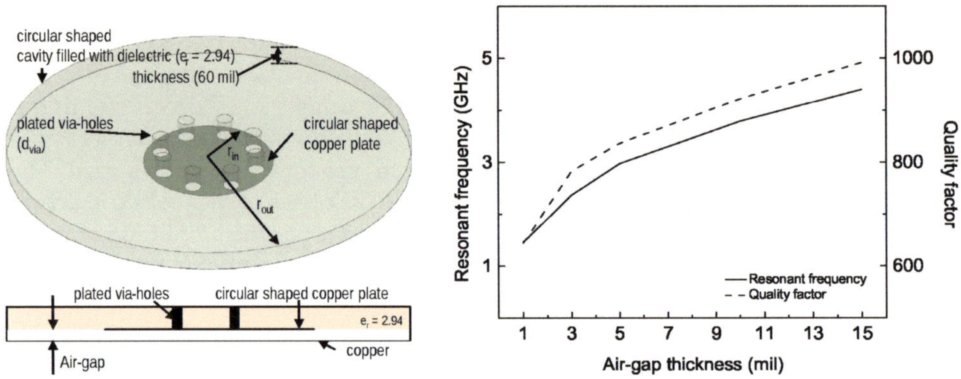

Figure 13.
Simplified view of a frequency tunable circular SIW resonator and its electrical characteristics.

plate and the copper membrane so the resonant frequency tuning mainly can be done by changing the thickness of an air gap. The relationship between the thickness and resonant frequency as well as the quality factor is simulated and the results are also given in **Figure 13**.

Based on the Eigen model simulation results, a single-frequency tunable resonator is designed to support the electromagnet-based tuning method. As shown in the upper left figure of **Figure 14**, two conductor-backed coplanar waveguide lines which are designed to have 50 Ω characteristic impedance are used as input and output lines. Two radial-shaped slots control the external couplings between 50 Ω line and resonator so a tight coupling can be achieved with a larger coupling slot. The side view is also given for the fabrication process, and it is noted that a high permeability foil is glued at the bottom of the resonator with the same adhesive used for the lamination of two substrates. The photograph of a fabricated resonator with a high permeability foil is also shown in **Figure 14**. The high permeability sheet glued to the frequency tunable SIW resonator for the magnetically actuated tuning method is from Metglas, Inc.

Prior to the implementation of the proposed frequency tuning method to the fabricated filter structure, the magnetic flux density and the current consumption from an electromagnet need to be investigated and the measured results are given in **Figure 15**. The magnetic flux density is measured using a Tesla meter and a probe as

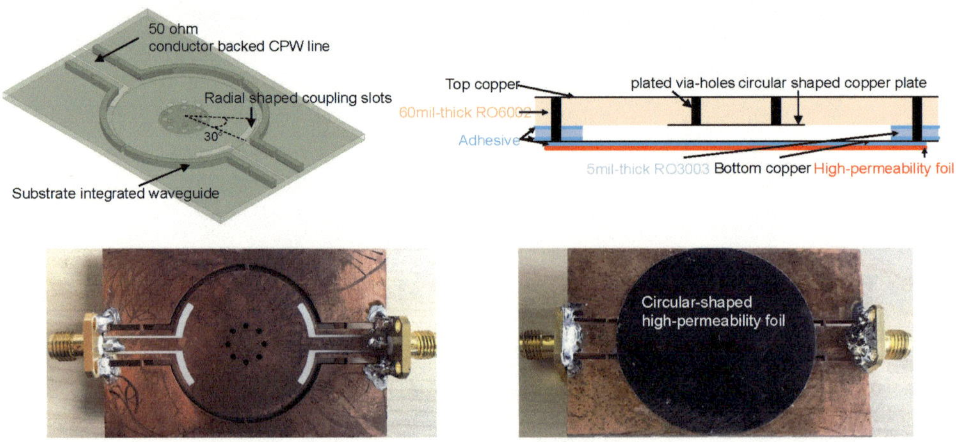

Figure 14.
Simulation model of frequency tunable SIW resonator, a side view drawing of the fabricated circuit, and photographs of SIW resonator.

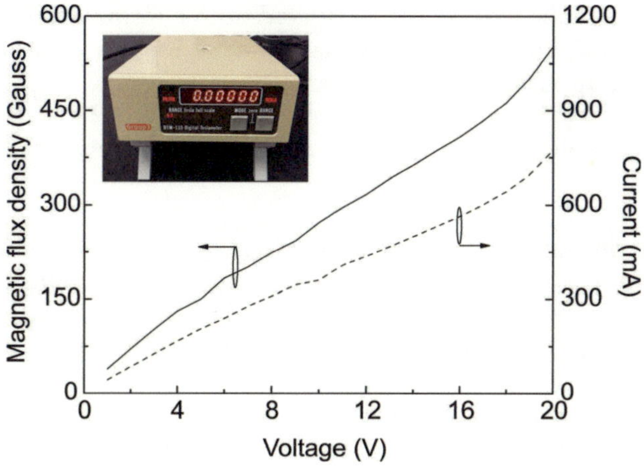

Figure 15.
Measured magnetic flux density and current consumption from an electromagnet.

shown in the inset of **Figure 15**. An electromagnet used for the measurement is a readily available one from market and its rated input voltage is 12 V, so applying the input voltage larger than the rated voltage for a long period of time can result in a damage of the electromagnet.

The electromagnet is placed both upper and lower sides of the fabricated one-pole frequency tunable SIW resonator to tune the resonant frequency as shown in **Figure 16**. Since the electromagnet only generates an attraction force from the input voltage, two electromagnets are used to move the high permeability foil in the opposite direction. In order to maximize the frequency tuning range, the input voltage to the electromagnet is applied to each at a time. The resonant frequency is tuned from 1.3 to 3.7 GHz which is larger than the frequency tuning ratio of 2.8:1.

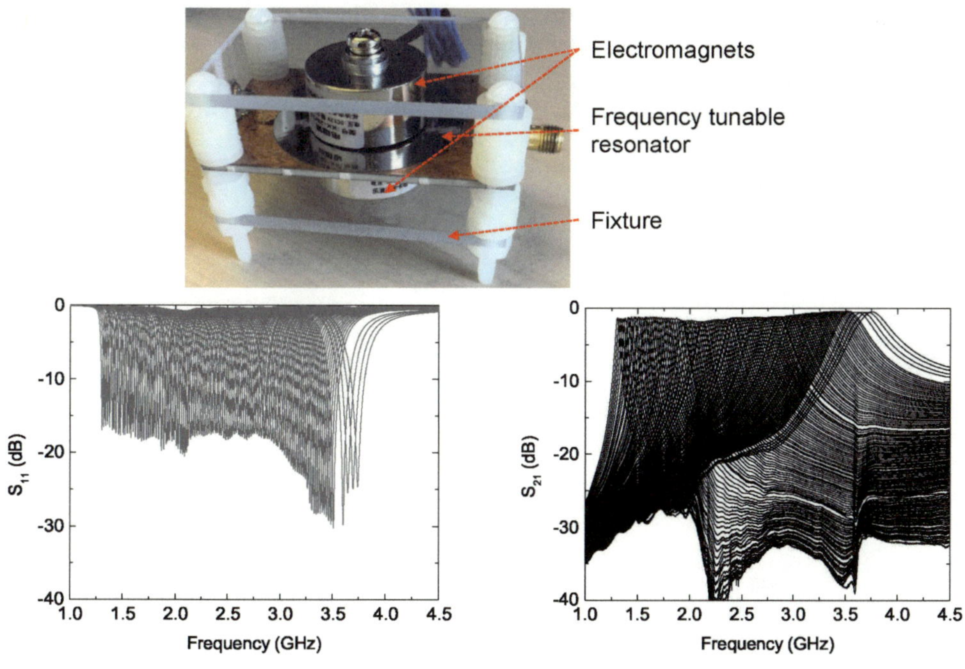

Figure 16.
Photograph of the fabricated one-pole frequency tunable SIW resonator with electromagnets and its measured frequency responses.

In future research on the magnetically actuated tuning elements, the smaller electromagnetic that can generate two-way magnetic forces with lower input voltage to satisfy the specific applications are expected.

3.3 Frequency tunable filtering balun with magnetically actuated tuning method

In this subsection, a filtering balun design is provided and its resonant frequency is tuned based on the aforementioned magnetically actuated tuning method. The fundamental design theory for the frequency tunable filtering balun follows the one reported in [9] except for the order of circuit structure and the required frequency tuning range.

Figure 17 shows the exploded view of the frequency tunable filtering balun. Similar to the reconfigurable filter given in [8], the filtering balun consists of two different substrates and each substrate contains the microstrip lines or SIW resonators, respectively. The couplings between two substrates are achieved through coupling slots placed at the ground plane of the microstrip line. To meet the requirement of balun, one port is designed in form of a single-ended microstrip input line and the other two ports have differential output lines. The short-circuited microstrip line fed the SIW resonator and the two ports connected to the other microstrip line receive output signal having an equal magnitude and a phase difference of 180°. This can be done by introducing a coupling slot at the center of the U-shaped microstrip [9]. As shown in the photograph of the fabricated filtering balun, two circular-shaped foils which have a high permeability are glued at the bottom of the devised circuit. Two electromagnets are placed at both sides of the circuit for each frequency tunable SIW resonator. To satisfy the required frequency tuning range, the electromagnets are placed as close as possible by optimizing the height of the plastic support. The electromagnets are the same as those used in **Figure 16**, so the input voltage can also be swept from 0 V to 12 V.

Figure 19 presents the simulation results of the filtering balun. The required frequency tuning is about 17% at the center frequency of 2.9 GHz with the proper performance on the differential output line such as the amplitude and phased imbalance. The mixed-mode S-parameters (S_{ds21}, S_{cs21}, and S_{dd22}) are calculated and

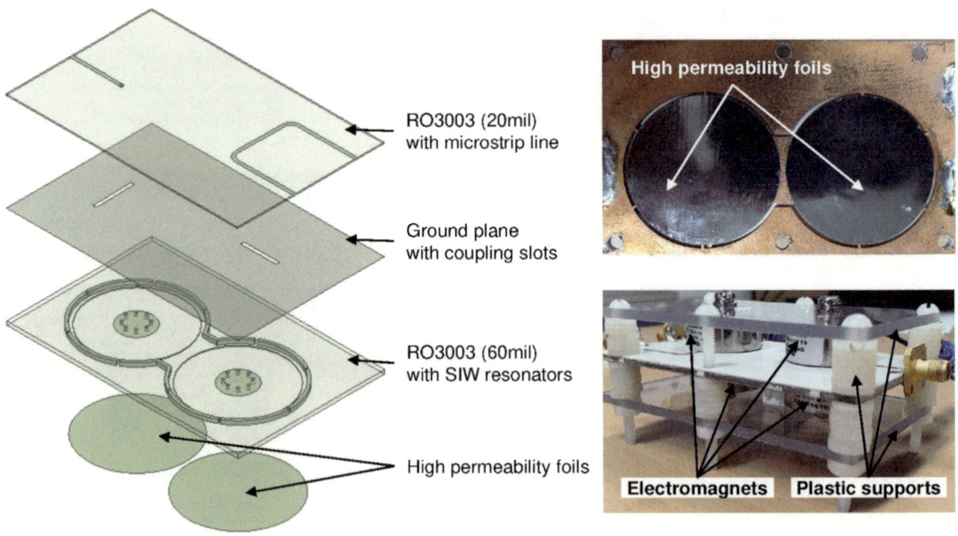

Figure 17.
Simulation model of the frequency tunable filtering balun and its detailed view with high permeability foils.

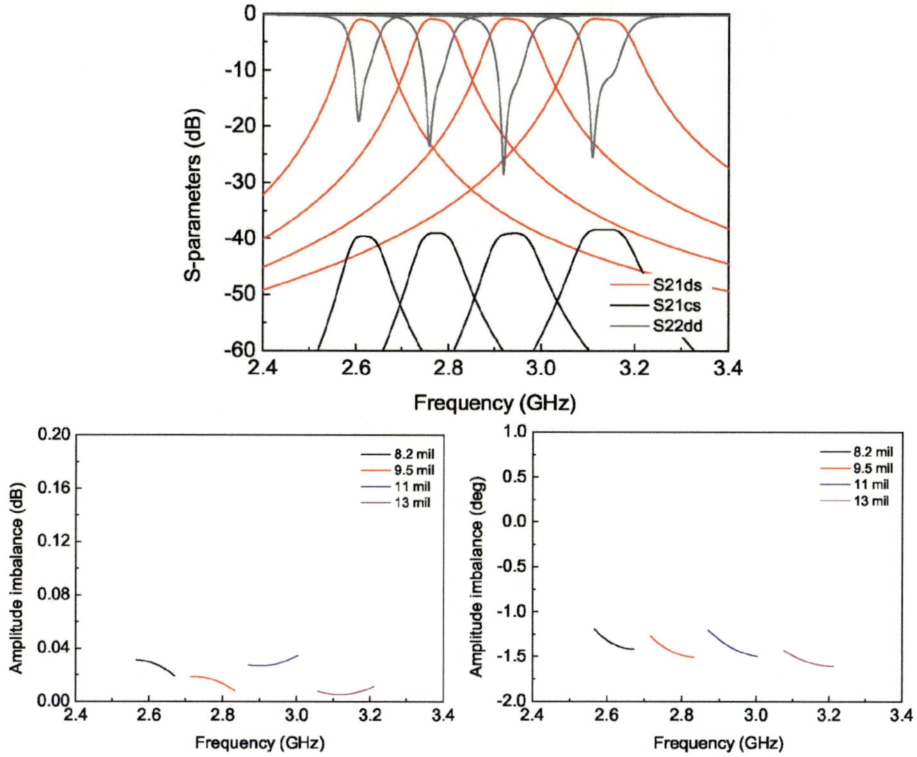

Figure 18.
Simulation results of the frequency tunable filtering balun.

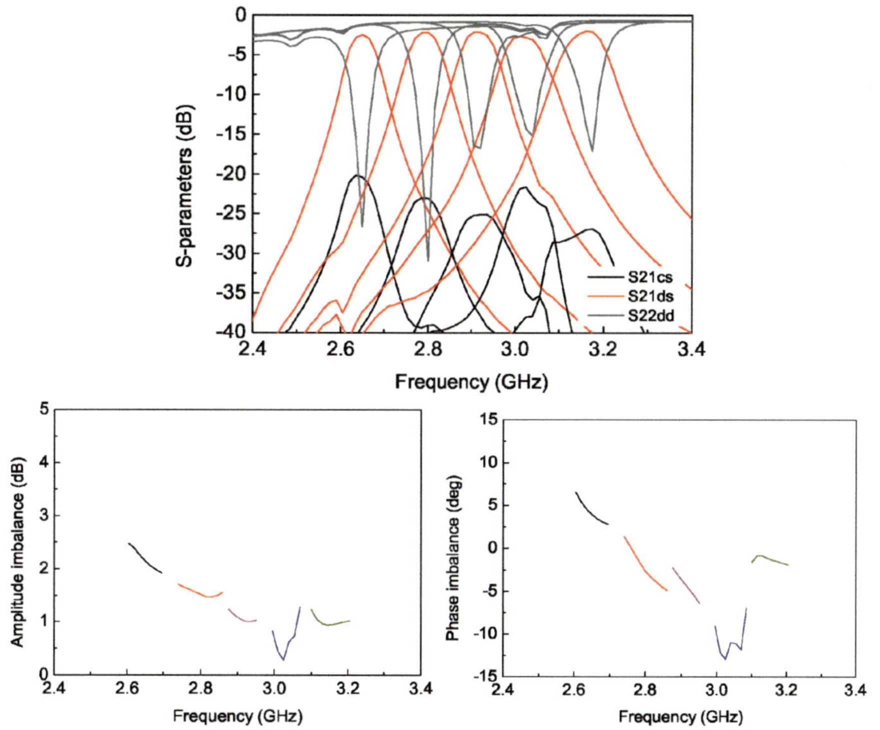

Figure 19.
Measurement results of the frequency tunable filtering balun.

also given in **Figures 18** and **19**. The operating frequency of the fabricated filtering balun is tuned from 2.65 to 3.15 GHz which satisfies the requirement and both amplitude and phase imbalanced performance at the passbands are given. Some discrepancies between the simulated and measured results are from some unexpected factors raised from fabrication or assemblies that can impact the electrical characteristic of the differential output signal. In this section, the frequency tuning method utilizing the electromagnets with high-permeability foil has been tested and the measurement results show that it can provide a comparable performance with the one with piezoelectric disks.

4. Conclusions

In this chapter, we explore the design process of the reconfigurable filter which can exhibit both frequency tunable bandpass and absorptive bandstop frequency responses. The coupling structures that can satisfy the predetermined requirements are designed from the theoretical normalized coupling coefficients, and its simulation/measurement models are also given. In order to realize the frequency tunable characteristic, two different tuning elements which are based on the piezoelectric disks or electromagnets are shown with the operation mechanism, and its application to the fabricated filtering balun is implemented especially using the electromagnets.

Acknowledgements

The authors want to thank Mr. Traian Antonescu and Mr. Steve Dubé from Ecole Polytechniqe, PolyGrames Research Center for their professional fabrication.

Author details

Tae-Hak Lee[1*], Sang-Gyu Lee[2], Jean-Jacques Laurin[3] and Ke Wu[3]

1 Yuhan University, Department of Electronic Engineering, Bucheon, Republic of Korea

2 Korea Aerospace Research Institute, Satellite Payload Research and Development Division, Daejeon, Republic of Korea

3 PolyGrames Research Center, École Polytechnique (Montréal University), Montréal, QC, Canada

*Address all correspondence to: taehaklee@gmail.com

References

[1] Cameron RJ, Kudsia CM, Mansour RR. Microwave Filters for Communication Systems: Fundamentals, Design, and Application. 2nd ed. Hoboken: Wiley; 2018. DOI: 10.1002/9781119292371

[2] Hong JS. Microstrip Filters for RF/ Microwave Applications. 2nd ed. New York: Wiley; 2011. DOI: 10.1002/ 9780470937297

[3] Matthaei G, Jones EMT, Young L. Microwave Filters, Impedance Matching Networks, and Coupling Structures. Norwood: Artech House; 1980

[4] Lee T-H, Laurin J-J, Wu K. Reconfigurable filter for bandpass-to-absorptive bandstop responses. IEEE Access. 2020;**8**:6484-6495. DOI: 10.1109/ACCESS.2019.2963710

[5] Yang T, Rebeiz GM. Bandpass-to-bandstop reconfigurable tunable filters with frequency and bandwidth controls. IEEE Transaction on Microwave Theory and Technology. 2017;**65**:2288-2297. DOI: 10.1109/TMTT.2017.2679182

[6] Moon S, Sigmarsson HH, Joshi H, Chappell WJ. Substrate integrated evanescent mode cavity filter with a 3.5 to 1 tuning ratio. IEEE Microwave Wireless Componts Letters. 2010;**20**: 450-452. DOI: 10.1109/ LMWC.2010.2050680

[7] Lee B, Nam S, Lee T-H, Ahn C-S, Lee J. Single-filter structure with tunable operating frequency in noncontiguous bands. IEEE Transactions on Components, Packaging and Manufacturing Technology. 2017;**7**:98-105. DOI: 10.1109/TCPMT.2016.2623804

[8] Lee T-H, Laurin J-J, Wu K. A wideband frequency-tuning method using magnetically actuated mechanical tuning of a SIW resonator. In: Proceedings of the IEEE International Microwave Conference (IMS 2019); 1-7 June 2019; Boston. MA: IEEE; 2019. pp. 1-4

[9] Hickle MD, Peroulis D. A widely-tunable substrate-integrated balun filter. In: Proceedings of the IEEE International Microwave Conference (IMS 2017); 4-9 June 2017; Honololu. HI: IEEE; 2017. pp. 1-4

Section 2

Kalman Filters and Analysis

Chapter 3

Kalman Filter Estimation and Its Implementation

Erick Ulin-Avila and Juan Ponce-Hernandez

Abstract

In this chapter, we use the Kalman filter to estimate the future state of a system. We present the theory, design, simulation, and implementation of the Kalman filter. We use as a case example the estimation of temperature using a Resistance Temperature Detector (RTD), which has not been reported before. After a brief literature review, the theoretical analysis of a Kalman filter is presented along with that of the RTD. The dynamics of the RTD system are analytically derived and identified using Matlab. Then, the design of a time-varying Kalman filter using Matlab is presented. The solution to the Riccati equation is used to estimate the future state. Then, we implement the design using C-code for a microprocessor ATMega328. We show under what conditions the system may be simplified. In our case, we reduced the order of the system to that of a system having a 1st order response, that of an RC system, giving us satisfactory results. Furthermore, we can find two first order systems whose response defines two boundaries inside which the evolution of a second order system remains.

Keywords: Kalman filter, prediction, Riccati equation

1. Introduction

A deterministic system is a system whose governing physical laws are specified so that if the state of the system at some time is known, then one can precisely predict the state at a later time. Nondeterministic systems are divided into two categories: stochastic and random. A stochastic system has governing physical laws that even if the state at some point in time is known precisely, it is impossible to determine the state of the system at a later time precisely. It is possible to determine the probability of a state, rather than the state itself. A random system is one which has no apparent governing physical laws. Practically, we treat all unpredictable systems, stochastic or random as stochastic systems, since we employ the same methods to study them. While we are unable to predict the state of a random process, we can evolve a strategy to deal with such processes. Such a strategy is based on a branch of mathematics dealing with unpredictable systems, called statistics.

Estimation is the process of extracting information from data which can be used to predict the behavior of state variables in a system. The estimation uses statistical criteria to infer the actual value of unknown variables. Estimation models are used to process noisy measurements, filter them, and detect inaccuracies. When random signals are passed through a deterministic system, their statistical properties are

IntechOpen

modified. A deterministic system to which random signals are input, so that the output is a random signal with desired statistical properties is called a filter. Filters can be linear or nonlinear, time-invariant or time varying. However, for simplicity we will usually consider linear, time-invariant filters. Linear, time-invariant filters are commonly employed in control systems to reduce the effect of measurement noise on the control system. In such systems, the output is usually a superposition of a deterministic signal and a random measurement noise.

The output of a filter not only has a frequency content different from the input signal, but also certain other characteristics of the filter, such as a phase-shift or a change in magnitude. In other words, the signal passing through a filter is also distorted by the filter, which is undesirable. A filter would produce an output signal based upon its characteristics, described by the transfer-function, frequency or impulse response, or a state-space representation of the filter. However, a filter can be designed to achieve a desired set of performance objectives, i.e. the numerator and denominator polynomials of the filter's transfer function, or coefficient matrices of the filter's state-space model, can be selected by a design process to achieve the conflicting requirements of maximum noise attenuation and minimum signal distortion.

There are several prediction models to infer the system state, although, it can be shown that of all estimation tools Kalman Filter (KF) is the one that minimizes the variance of the estimation error which enables accurate estimation of the process.

1.1 Literature review

The first application of state estimation was in the aerospace field to solve problems related to the prediction of position in aerospace vehicles. Nowadays, estimation has been applied in several fields of engineering and control systems. One common application is in data acquisition, to solve the problem of predicting the state of a system that cannot be measured directly due to the characteristics and complexity of the environment.

KF is an estimator proposed by Rudolph E. Kalman in 1960. It is an algorithm to estimate the evolution of a dynamic system, especially when data has a lot of noise. The principle of the filter is to find the probability of the hypothesis of predicted state and using the data from the measurement to correct it and improve the future estimation at each time. It is a suitable algorithm to apply in dynamic systems, linking real-time measurements and predicting the state of system parameters through time approaches. KF has been implemented in several fields, such as in navigation systems [1–4], financial models [5–7], tracking vehicles [8, 9] and image processing [10–12]; only to mention some of them. Nevertheless, this statistical tool is useful for two main purposes: estimation and performance analysis of estimators.

In the field of IC technology, it has been implemented for thermal estimation. Multicore processors use a dynamic thermal management mechanism that use embedded thermal sensors for monitoring the real-time thermal behavior of the processor, this kind of sensors are susceptible to a variety of source of noise and this causes the discrepancies between actual temperatures observed by on-chip thermal sensors. Therefore, to fix the discrepancies in sensing, Kalman's prediction is used to estimate real values from noisy sensor readings [13]. Another novel application of KF is in the electric vehicle industry, the estimation of the charge state of lithium-ion battery is an important parameter in order to guarantee a safe operation of them. The battery performance is influenced by aging; this fact makes difficult to predict the battery state, to overcome this issue the application of KF in combination with other methods is a suitable methodology [14–17].

Recently, KF has been applied in several industrial applications. With the development of manufacturing process, welding automation emerges as one important tool to speed up the production rate in the assembly line in stronger and high-quality welds. Nevertheless, there are several factors that could influence the welding quality and the most important is the arc length, which could be influenced by the irregular surface of the workpiece and the loss of the tungsten electrode. To enhance the quality during the Gas-Tungsten Arc Welding (GTAW) process, KF is applied in order to keep the arc length stable and minimize the external noise [18]. In the field of sensorless control, KF have been used in intelligence electrical drives. To control induction motor drives without mechanical speed sensor at the motor shaft allows reduced hardware complexity, and low costs. Additionally, the use of induction motors without position sensor is useful for applications with abrasive and hard surface. Thereby, the application of an estimation method it's necessary in order to predict the position and velocity of the shaft [19–21].

In applications related with radio astronomy, KF has been applied for the analysis of Very-Long-Baseline Interferometry (VLBI) data, in order to analyze parameters such as base line lengths, earth orientation parameters, radio source coordinates and tropospheric delays. Nowadays, modern antennas are being constructed and equipped with highly accurate broadband receiving systems. Besides the accurate observations gotten by astronomic instruments, it is necessary to implement estimation methods in order to optimize the models applied in data analysis [22, 23]. In power systems, one of the main difficulties is power quality due to total harmonics distortion (THD) that is mainly caused by nonlinear loads. THD effects are strongly correlated with issues as device heating, break down electronic components, network interference, etc. Several filters have been performed to decrease the effect of harmonics; nevertheless, the application of KF has shown an important reduction in the effect of harmonics [24–26]. In the field of biomedicine KF is widely used over other estimation methodologies to overcome the different sources of noise. Specifically, KF has been used to smooth and predict signals from Electroencephalogram and Electrocardiogram signals [27, 28]. Recently in the literature there are reports on a new methodology to protect the confidentiality of the transmitted data based on a Kalman filter. This strategy proposes the implementation of encrypted algorithm using KF, and is suggested to be used in Industrial cyber–physical systems (ICPSs) to protected data privacy [29, 30].

As it has been mentioned above, KF has been used in diverse fields of science and technology to predict specific parameters of interest according to the application. Temperature evolution is an important parameter to measure and predict, in order to study or control the temperature in an environment [31, 32], device [13, 33, 34] and process [18]. It is well known that RTDs are commercial devices very useful to monitor the temperature due their stability and accuracy. However, RTDs are self-heating causing noisy readings making the RTD a suitable example to implement KF for temperature estimation. Importantly, we searched in the literature and found no evidence of previous work reporting the use of a KF to filter the noise and predict the temperature behavior from RTD readings.

2. Theoretical analysis of a Kalman Filter

The final objective of this study is to obtain the specification of a linear dynamic system (Wiener filter [35]) which accomplishes the prediction, separation, or detection of a random signal [36]. With the state-transition method, a single derivation covers a large variety of problems: growing and infinite memory filters, stationary and non-stationary statistics, etc. Having guessed the "state" of the estimation (i.e., filtering or prediction) problem correctly, one is led to a nonlinear difference (or differential) equation for the covariance matrix of the optimal estimation error. From the solution of the equation for the covariance matrix we obtain the coefficients characterizing the optimal linear filter [36]. The following is a simplified derivation described previously in the references [37, 38].

2.1 Defining statistical quantities of use

The initial state, $\mathbf{x}(0)$, of a stochastic system is insufficient to determine its future state, x(t). Thus, based upon a statistical analysis of similar systems, and taking the average of their future states at a given time, t, we can calculate the mean state-vector as follows:

$$\mathbf{x}_{\mathrm{m}}(\mathrm{t}) = (1/\mathrm{N}) \sum_{\mathrm{i}=1}^{\mathrm{N}} \mathrm{x}_{\mathrm{i}}(\mathrm{t}) \tag{1}$$

Thus $\boldsymbol{x}_m(t)$ is the expected state vector after studying N systems. It is also called the expected value of the state-vector, $\boldsymbol{x}_m(t) = E[\boldsymbol{x}(t)]$. Another statistical quantity of use is the correlation matrix of the state-vector:

$$\mathbf{P}_x(\mathrm{t}, \tau) = (1/N) \sum_{i=1}^{N} x_i(t) x_i^T(\tau) \tag{2}$$

The correlation matrix, $\mathbf{P}_x(\mathrm{t}, \tau)$, is a measure of correlation, a statistical property among the different state variables, and between the same state variable at two different times. Two scalar variables, $x_1(t)$ and $x_2(t)$, are said to be uncorrelated if the expected value of $x_1(t)x_2(\tau)$, i.e. $E[x_1(t)x_2(\tau)] = 0$, where τ is different from t.

The correlation matrix is the expected value of the matrix $\boldsymbol{x}_i(t)\boldsymbol{x}_i^T(\tau)$, or $\mathbf{P}_x(\mathrm{t}, \tau) = \mathrm{E}\left[x_i(t)x_i^T(\tau)\right]$. When t = τ, the correlation matrix $\mathbf{P}_x(\mathrm{t}, \mathrm{t}) = \mathrm{E}\left[x_i(t)x_i^T(t)\right]$, is called the covariance matrix. The covariance matrix, $\mathbf{P}_x(\mathrm{t}, \mathrm{t})$, is symmetric. If $\mathbf{P}_x(\mathrm{t}, \tau)$ is a diagonal matrix i.e. $E\left[x_i(t)x_j(\tau)\right] = 0$, where $\mathrm{i} \neq \mathrm{j}$, it implies that all the state variables are uncorrelated.

2.2 Defining the filter in state space - discrete domain

Consider a plant which we cannot model accurately using only a deterministic model, because of the presence of uncertainties called process noise and measurement noise:

$$\boldsymbol{x}_{k+1} = \boldsymbol{A}\boldsymbol{x}_k + \boldsymbol{w}_k \tag{3}$$

$$\boldsymbol{y}_k = \boldsymbol{C}\boldsymbol{x}_k + \boldsymbol{v}_k \tag{4}$$

In the linear, time-varying state-space representation above, $\mathbf{w}$ is the process noise vector which may arise due to modeling errors such as neglecting nonlinear dynamics, and $\mathbf{v}$ is the measurement noise vector. The random noises, $\mathbf{w}$ and $\mathbf{v}$, are assumed to be stationary white noises. The covariance matrices of stationary white noises, $\mathbf{w}$ and $\mathbf{v}$, can be expressed as follows:

$$\mathbf{Q} = \mathrm{E}\left[w_k w_k^T\right] \tag{5}$$

$$\mathbf{R} = \mathrm{E}\left[v_k v_k^T\right] \tag{6}$$

Since we cannot predict the state-vector, $\mathbf{x}$ of a stochastic plant, an observer is required for estimating the state-vector, based upon a measurement of the output, $\mathbf{y}$ and a known input, $\mathbf{u}$. We need an observer that calculates the estimated state-vector, $\hat{\mathbf{x}}$, optimally, based upon statistical description of the vector output and plant state. Such an observer is the Kalman Filter, which minimizes a statistical measure of the estimation error, $\mathbf{e}_k = \mathbf{x}_k - \hat{\mathbf{x}}_k$. This statistical measure is the covariance of the estimation error:

$$\mathbf{P}_k = \mathrm{E}\left[\boldsymbol{e}_k \boldsymbol{e}_k^T\right] = E\left[(\mathbf{x}_k - \hat{\mathbf{x}}_k)\,(\mathbf{x}_k - \hat{\mathbf{x}}_k)^T\right] \tag{7}$$

Since the state-vector, $\mathbf{x}$, is a random vector and the estimated state $\hat{\mathbf{x}}$, is based on the measurement of the output, $\mathbf{y}$, for a finite time, say T, where $t \geq T$ then a true statistical average of $\mathbf{x}$ would require measuring the output for an infinite time.

If T < t, this is a data-smoothing (interpolation) problem. If T = t, this is called filtering. If T > t, we have a prediction problem. Since the original treatment is general enough, the collective term estimation is used [36].

Hence, the best estimate to obtain for $\mathbf{x}$ is not the true mean, but a conditional mean, $\boldsymbol{x}_m$, based on only a finite time record of the output, $\mathbf{y}$:

$$\boldsymbol{x}_m = E\left[\boldsymbol{x} : \boldsymbol{y}, T \leq t\right] \tag{8}$$

Taking in consideration the deviation of the estimated state- vector, $\hat{\mathbf{x}}$, from the conditional mean, $\boldsymbol{x}_m$, we can write the estimated state- vector as:

$$\hat{\mathbf{x}} = \boldsymbol{x}_m + \Delta\boldsymbol{x} \tag{9}$$

$\Delta\boldsymbol{x}$ is the deviation from the conditional mean. The conditional covariance matrix of the estimation error based on a finite record of the output is then:

$$\mathbf{P}_k = \mathrm{E}\left[\boldsymbol{e}_k \boldsymbol{e}_k^T : \boldsymbol{y}, T \leq t\right] = E\left[\boldsymbol{x}\boldsymbol{x}^T\right] - \boldsymbol{x}_m \boldsymbol{x}_m^T + \Delta\boldsymbol{x}\Delta\boldsymbol{x}^T \tag{10}$$

The best estimate of state-vector happens if $\Delta\boldsymbol{x} = 0$, or $\hat{\mathbf{x}} = \boldsymbol{x}_m$, and would result in a minimization of the conditional covariance matrix, or error covariance matrix, $\mathbf{P}_k$. In other words, minimization of $\mathbf{P}_k$ yields the optimal observer, which is the Kalman filter.

2.3 Defining the Kalman gain

The state-equation of the Kalman filter is that of a time-varying observer, and can be written as follows:

$$\hat{\mathbf{x}}_{k+1} = \boldsymbol{A}\hat{\mathbf{x}}_k + \boldsymbol{B}\boldsymbol{u}_k + \boldsymbol{K}_k\left[\boldsymbol{y}_k - \boldsymbol{C}\hat{\mathbf{x}}_k\right] \tag{11}$$

$\boldsymbol{K}_k$ is the gain matrix of the Kalman filter. Assuming the prior estimate of $\hat{\mathbf{x}}_k$ is called $\hat{\mathbf{x}}_k'$, gained by knowledge of the system. We write an update equation for the new estimate, combing the old estimate with measurement data, $\hat{\mathbf{x}}_k' = \boldsymbol{A}\hat{\mathbf{x}}_k + \boldsymbol{B}\boldsymbol{u}_k$ and:

$$\hat{\mathbf{x}}_k = \hat{\mathbf{x}}_k' + \boldsymbol{K}_k\left[\boldsymbol{y}_k - \boldsymbol{C}\hat{\mathbf{x}}_k'\right] \tag{12}$$

If we substitute Eq. (4) into Eq. (12) we get:

$$\hat{\mathbf{x}}_k = \hat{\mathbf{x}}'_k + \boldsymbol{K}_k\left[\,\boldsymbol{C}\boldsymbol{x}_k + \boldsymbol{v}_k - \boldsymbol{C}\hat{\mathbf{x}}'_k\right] \tag{13}$$

Substituting Eq. (13) into Eq. (7)

$$\mathbf{P}_k = E\left[\left((\boldsymbol{I} - \boldsymbol{K}_k\boldsymbol{C})\left(\mathbf{x}_k - \hat{\mathbf{x}}'_k\right) - \boldsymbol{K}_k\boldsymbol{v}_k\right)\left((\boldsymbol{I} - \boldsymbol{K}_k\boldsymbol{C})\left(\mathbf{x}_k - \hat{\mathbf{x}}'_k\right) - \boldsymbol{K}_k\boldsymbol{v}_k\right)^T\right] \tag{14}$$

Here $\left(\mathbf{x}_k - \hat{\mathbf{x}}'_k\right)$ is the error of the prior estimate. Since there is no correlation among the input, process noise and measurement noise, then the expectation may be re-written as;

$$\mathbf{P}_k = (\boldsymbol{I} - \boldsymbol{K}_k\boldsymbol{C})E\left[\left(\mathbf{x}_k - \hat{\mathbf{x}}'_k\right)\left(\mathbf{x}_k - \hat{\mathbf{x}}'_k\right)^T\right](\boldsymbol{I} - \boldsymbol{K}_k\boldsymbol{C})^T + \boldsymbol{K}_k\boldsymbol{E}\left[\boldsymbol{v}_k{\boldsymbol{v}_k}^T\right]\boldsymbol{K}_k^T \tag{15}$$

Using Eqs. (6) and (7), we obtain:

$$\mathbf{P}_k = (\boldsymbol{I} - \boldsymbol{K}_k\boldsymbol{C})\boldsymbol{P}'_k(\boldsymbol{I} - \boldsymbol{K}_k\boldsymbol{C})^T + \boldsymbol{K}_k\boldsymbol{R}\boldsymbol{K}_k^T \tag{16}$$

Eq. (16) is the error covariance update equation, where $\boldsymbol{P}'_k$ is the prior estimate of $\mathbf{P}_k$.

The trace of the error covariance matrix is the sum of the mean squared errors. The mean squared error may be reduced by minimizing the trace of $\mathbf{P}_k$. This requires to differentiate the trace of $\mathbf{P}_k$ with respect to $\boldsymbol{K}_k$, then the result set to zero to find $\boldsymbol{K}_k$ that minimizes the trace of $\mathbf{P}_k$.

We rewrite Eq. (16);

$$\mathbf{P}_k = \boldsymbol{P}'_k - \boldsymbol{P}'_k\boldsymbol{C}^T\boldsymbol{K}_k^T - \boldsymbol{K}_k\boldsymbol{C}\boldsymbol{P}'_k + \boldsymbol{K}_k\boldsymbol{C}\boldsymbol{P}'_k\boldsymbol{C}^T\boldsymbol{K}_k^T + \boldsymbol{K}_k\boldsymbol{R}\boldsymbol{K}_k^T \tag{17}$$

Taking the trace of this expression gives:

$$\mathbf{T}[\mathbf{P}_k] = \mathrm{T}\left[\boldsymbol{P}'_k\right] - 2\mathrm{T}\left[\boldsymbol{K}_k\boldsymbol{C}\boldsymbol{P}'_k\right] + T\left(\boldsymbol{K}_k\left(\boldsymbol{C}\boldsymbol{P}'_k\boldsymbol{C}^T + \boldsymbol{R}\right)\boldsymbol{K}_k^T\right) \tag{18}$$

Then, we differentiate with respect to $\boldsymbol{K}_k$;

$$\frac{\mathbf{dT}[\mathbf{P}_k]}{\mathrm{d}\boldsymbol{K}_k} = -2\mathrm{T}\left[\boldsymbol{C}\boldsymbol{P}'_k\right] + 2T\left(\boldsymbol{K}_k\left(\boldsymbol{C}\boldsymbol{P}'_k\boldsymbol{C}^T + \boldsymbol{R}\right)\right) \tag{19}$$

Setting to zero and solving for $\boldsymbol{K}_k$ we obtain the Kalman gain equation:

$$\boldsymbol{K}_k = \boldsymbol{P}'_k\boldsymbol{C}^T\left(\boldsymbol{C}\boldsymbol{P}'_k\boldsymbol{C}^T + \boldsymbol{R}\right)^{-1} \tag{20}$$

Substitution of Eq. (20) into [17], gives:

$$\mathbf{P}_k = (\boldsymbol{I} - \boldsymbol{K}_k\boldsymbol{C})\boldsymbol{P}'_k \tag{21}$$

Eq. (21) is the update equation for the error covariance matrix with optimal gain.

State projection is derived using;

$$\hat{\mathbf{x}}'_{\mathrm{k}+1} = \mathbf{A}\hat{\mathbf{x}}'_{\mathrm{k}} + \mathbf{w}_{\mathrm{k}} \tag{22}$$

To project the error covariance matrix into the next time interval, k + 1 we first find an expression for the error based on the prior error;

$$\begin{aligned}\mathbf{e}'_{\mathbf{k+1}} &= \mathbf{x}_{\mathrm{k+1}} - \hat{\mathbf{x}}'_{\mathrm{k+1}} \\ &= \mathbf{A}\mathbf{x}_{\mathrm{k}} + \mathbf{w}_{\mathrm{k}} - \mathbf{A}\hat{\mathbf{x}}'_{\mathrm{k}} \\ &= \mathbf{A}\mathbf{e}_{\mathrm{k}} + \mathbf{w}_{\mathrm{k}}\end{aligned} \tag{23}$$

Eq. (7) in time k + 1 is;

$$\mathbf{P}'_{k+1} = \mathrm{E}\left[e'_{k+1}{e'_{k+1}}^{T}\right] = E\left[(\boldsymbol{Ae_k} + \boldsymbol{w_k})(\boldsymbol{Ae_k} + \boldsymbol{w_k})^{T}\right] \tag{24}$$

Assuming that $\boldsymbol{e_k}$ and w_k have zero cross-correlation.

$$\begin{aligned}\mathbf{P}'_{k+1} &= \mathrm{E}\left[e'_{k+1}{e'_{k+1}}^{T}\right] \\ &= E\left[\left(\boldsymbol{A}\boldsymbol{e_k}\boldsymbol{e_k}^{T}\boldsymbol{A}^{T} + \boldsymbol{w_k}\boldsymbol{w_k}^{T}\right)\right] \\ &= \mathbf{A}\mathbf{P}_{\mathrm{k}}\mathbf{A}^{\mathrm{T}} + \mathbf{Q}\end{aligned} \tag{25}$$

This completes the description of the filter.

2.4 Algorithm loop

An algorithm loop is required to make the program in MATLAB and in C-code for the microprocessor. The loop is summarized in the **Figure 1**.

The KF assumes that the system model is linear and known, the system and measurement noises are white, and the states have initial conditions with known means and variances. The power spectral densities used can be treated as tuning parameters to design an observer with excellent performance and robustness. The linear Kalman filter can also be used to design observers for nonlinear plants, by treating nonlinearities as process noise with appropriate power spectral density matrix.

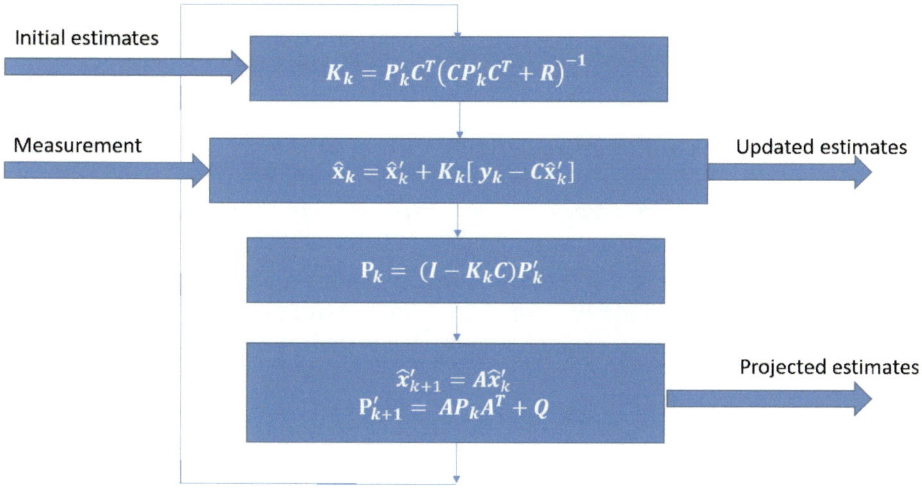

Figure 1.
Recursive algorithm for the Kalman filter.

2.5 Derivation of the Riccati equation

Since the Kalman filter is an optimal observer the appearance of matrix Riccati equation is not surprising. We are interested in a steady Kalman filter, i.e. the Kalman filter for which the covariance matrix converges to a constant in the limit $t \to \infty$. This happens when the plant is time invariant. The derivation goes as follows [39, 40]:

From the projections into ∞ we get:

$$\hat{\mathbf{x}}'_{\infty+1} = \mathbf{A}\hat{\mathbf{x}}_{\infty} \tag{26}$$

$$\mathbf{P}_{\infty+1} = \mathbf{A}\mathbf{P}_{\infty}\mathbf{A}^{T} + \mathbf{Q} \tag{27}$$

$$\mathbf{P}_{\infty} = (I - K_{\infty}C)P'_{\infty} \tag{28}$$

$$\hat{\mathbf{x}}_{\infty} = \hat{\mathbf{x}}'_{\infty} + K_{\infty}\left[y_{\infty} - C\hat{\mathbf{x}}'_{\infty}\right] \tag{29}$$

$$K_{\infty} = P'_{\infty}C^{T}\left(CP'_{\infty}C^{T} + R\right)^{-1} \tag{30}$$

Using the Eqs. 26–29 we get:

$$\hat{x}'_{\infty+1} = A\hat{\mathbf{x}}'_{\infty} + AK_{\infty}\left[y_{\infty} - C\hat{\mathbf{x}}'_{\infty}\right] \tag{31}$$

$$P_{\infty+1} = A(I - K_{\infty}C)P'_{\infty}A^{T} + Q \tag{32}$$

Using Eq. 30 in Eqs. 31 and 32 we get:

$$\hat{x}'_{\infty+1} = A\hat{\mathbf{x}}'_{\infty} + AP'_{\infty}C^{T}\left(CP'_{\infty}C^{T} + R\right)^{-1}\left[y_{\infty} - C\hat{\mathbf{x}}'_{\infty}\right] \tag{33}$$

$$P_{\infty+1} = A\left(I - P'_{\infty}C^{T}\left(CP'_{\infty}C^{T} + R\right)^{-1}C\right)P'_{\infty}A^{T} + Q \tag{34}$$

Rewriting Eq. 34 we get:

$$P_{\infty+1} = AP'_{\infty}A^{T} - AP'_{\infty}C^{T}\left(CP'_{\infty}C^{T} + R\right)^{-1}CP'_{\infty}A^{T} + Q \tag{35}$$

When in steady state:

$$P_{\infty+1} = P'_{\infty} = P_{\infty} \tag{36}$$

Then we arrive at the Riccati equation:

$$AP_{\infty}A^{T} - AP_{\infty}C^{T}\left(CP_{\infty}C^{T} + R\right)^{-1}CP_{\infty}A^{T} - P_{\infty} + Q = 0 \tag{37}$$

The iterative solution of the Riccati equation is not required in real time. The observer gain is calculated off-line for predictive control applications [40]. Riccati equations are mainly used to control large scale systems, estimation, and, detection processes.

2.6 Solution to the Riccati equation using MATLAB

In this work the discrete-time algebraic Riccati equation (DARE) was solved to obtain the covariance matrix P of the Kalman gain. The discrete-time algebraic Riccati equation is represented by the next form [41]:

$$X = A^T XA + Q - (A^T XB)(R + B^T XB)^{-1}(B^T XA) \tag{38}$$

Where $A, X, Q = A^T \in \mathbb{R}^{n\times n}$, $B \in \mathbb{R}^{n\times m}$, $R \in \mathbb{R}^{m\times m} (m \le n)$, and $R = B^T > 0$. Eq. (38) can be written in the short form:

$$A^T X (I + SX)^{-1} A - X + Q = 0 \tag{39}$$

Where:

$$S = BR^{-1}B^T \tag{40}$$

The application of the Kalman filter implies solving the DARE, which can be solved by several solution methods. Computational methods to solve Riccati equations can be categorized into three classes: invariant subspace methods, deflating subspace methods, and Newton's methods. The generalized Schur method that is classified as a deflating subspace method is used to solve DARE. The generalized Schur algorithm is a strong algebraic tool that allows computing classical decompositions of matrices, such as the QR and LU factorizations [42]. The next algorithm was used to solve DARE [43]:

Input arguments:

$$\begin{aligned} &\mathrm{A} - \mathrm{An\ n \times n\ matrix} \\ &\mathrm{B} - \mathrm{An\ n \times m\ matrix} \\ &\mathrm{Q} - \mathrm{An\ n \times n\ symetric\ matrix} \\ &\mathrm{R} - \mathrm{An\ m \times m\ symetrix\ matrix} \end{aligned}$$

Output arguments: X – *DARE solution*

1. Form the pencil $P_{DARE} - \lambda N_{DARE}$, where

$$P_{DARE} = \begin{pmatrix} A & 0 \\ -Q & I \end{pmatrix}, \tag{41}$$

$$N_{DARE} = \begin{pmatrix} I & S \\ 0 & A^T \end{pmatrix} \tag{42}$$

2. Transform the pencil $P_{DARE} - \lambda N_{DARE}$ to the generalized real Schur form apply QZ algorithm, that is, find orthogonal matrices Q_1 and Z_1 such that:

$$Q_1 P_{DARE} Z_1 = P_1 = \begin{pmatrix} P_{11} & P_{12} \\ 0 & P_{22} \end{pmatrix}, \tag{43}$$

$$Q_1 N_{DARE} Z_1 = N_1 = \begin{pmatrix} N_{11} & N_{12} \\ 0 & N_{22} \end{pmatrix} \tag{44}$$

3. Using an orthogonal transformation and reorder the generalized real Schur form. So that all the pencil $P_{11} - \lambda N_{11}$ has all the eigenvalues with moduli less than 1. Find Q_z and Z_2 orthogonal matrices, such that:

$$Q_2Q_1P_{DARE}Z_1Z_2 = quasi - upper\ triangular \tag{45}$$

$$Q_2Q_1N_{DARE}Z_1Z_2 = upper\ triangular \tag{46}$$

4. Form the matrix:

$$Z = Z_1Z_2 = \begin{pmatrix} Z_{11} & Z_{12} \\ Z_{21} & Z_{22} \end{pmatrix} \tag{47}$$

5. *Compute* $X = Z_{21}Z_{11}^{-1}$

3. Application example: resistive temperature detectors (RTD)

Resistive temperature detectors (RTD) have attracted attention to be employed as thermal health monitors. As clinical thermometers they are stable and reliable presenting high accuracy and resolution [44]. One of the most widely used RTD is the emerging thin-film resistor which has minimal impact on complex circuits due to its small size and due to their negligible mass.

The basic function of the sensor is determined by a proportional increment of resistance when temperature is applied. RTDs can be employed on a rigid or flexible substrate [45–47], the metal combination with a flexible o rigid substrate can cover conformal applications. RTD fabrication can be done by metals like Pt [48–50], Cu [51], Ag [52], and Ni [53], among other materials. Nickel presents a suitable option for RTD fabrication due to its wide temperature linear range of operation and its relatively low price.

Clinical thermometers require a high definition and reliability because less than 1°C difference can indicate a health problem. The thermometer signal can be amplified by electronic means, but it is desirable to filter such readings. This work is focused to the filtering and prediction of an highly sensitive Nickel based thin film RTD (range, 273–325 K), to be incorporated to complex circuits [54], we present the theoretical analysis about the relation sensibility-resistance that matches with experimental results.

3.1 Theoretical analysis of an RTD

All metals produce an increase in its resistance to an increase in specific temperature, which means that resistance is linearly proportional to temperature change. This dependence between electrical resistance and temperature is the principle of operation used by a resistance temperature detector (RTD). The relation between temperature-resistance for Pt wire (RTD), is described by the equation known as the Calendar-Van Dusen, Eq. 41) [50].

$$R_{(T)} = R_{0°C}\left(1 + \alpha T + \beta T^2\right) \tag{48}$$

Where $R_{0°C}$ is the resistance at 0°C, α and β are temperature coefficients and T is temperature, the temperature coefficients depend only on material properties. In addition, the RTD resistance depends on its geometrical design, according to Eq. 42.

$$R = \frac{\sigma * L}{A} = \frac{\sigma * L}{w * t} \tag{49}$$

Where "σ" is the resistivity, "L" length, "A" lateral area, "w" channel width, and "t" channel height. Only by increasing the length "L" or decreasing the area "A" that means reducing the "t" film thickness or the "w" channel wide, the resistance can increase.

3.1.1 State-space description of an RTD

The estimation of the thermal system is represented by the linear stochastic state-space description $x_k = Ax_{k-1} + Bu_{k-1} + w_{k-1}$ and $y_k = Cx_k + v_k$. Where A is an nxn state transition matrix applied to the previous state vector x_{k-1}, B is the control-input matrix applied to the control vector u_{k-1}, and w_{k-1} is the process noise vector. The linear combination of the measurement noise and the signal value is represented by y_k, where C is the measurement matrix, and v_k is the measurement noise vector with covariances matrices represented by Q and R. The covariances are assumed to be independent and are given by $Q = E\left[w_k w_k^T\right]$ and $= E\left[v_k v_k^T\right]$.

Generally, the RTD system is modeled as an RLC circuit, which consists of a resistor a capacitor and an inductor in series with an input voltage. The output that we analyzed is the voltage across the resistor which is related to temperature change. The RLC circuit is represented by a second-order differential equation $L\frac{d^2i(t)}{dt^2} + R\frac{di(t)}{dt} + \frac{1}{C}i(t) = 0$. To solve the above equation we implement the next matrix system:

$$\mathrm{A} = \begin{pmatrix} 0 & 1 \\ -1/\mathrm{L}\cdot\mathrm{C} & -\mathrm{R}/\mathrm{L} \end{pmatrix} \qquad \mathrm{B} = \begin{pmatrix} 0 \\ 1/\mathrm{L}\cdot\mathrm{C} \end{pmatrix} \qquad (50)$$
$$\mathrm{C} = (1\ \ 0) \qquad \mathrm{D} = 0$$

Also, we may simplify the response of the system to that of a first-order RC circuit. This implies to solve a first-order ordinary differential equation: $RC\frac{dq}{dt} + q = VC$. The dynamic model is defined by the following system:

$$\mathrm{A} = \left(-\frac{1}{\mathrm{R}\cdot\mathrm{C}}\right) \qquad \mathrm{B} = \left(\frac{1}{\mathrm{R}\cdot\mathrm{C}}\right) \qquad (51)$$
$$\mathrm{C} = 1 \qquad \mathrm{D} = 0$$

4. Design and simulations

4.1 Kalman filter in resistance thermal detectors (RTD)

In this work, a Kalman Filter is proposed to decrease the time response to improve the speed feedback and filtering of the perturbations by signal noise from physical signals as thermal detectors. In some instances, a reduced model is advisable to use in an embedded system due to easy implementation and low computational complexity [2].

Kalman filter can be embedded in a temperature system made by Resistance Thermal Detectors (RTD).RTD's are robust elements that require relatively easy measurement, as a consequence are a useful thermal sensor for industry and medical applications. Nevertheless, these devices are exposing to vibration, electrical noise, and measurement errors generated by the thermoelectric effect caused by the temperature difference between electrical contacts, which affects the response time of the sensor. The implementation of the Kalman filter in a temperature system produces an optimal estimative of thermal behavior and decreases the uncertainties about the prediction of the temperature.

In order to describe the system in the state space, it is necessary to apply system identification methods using MATLAB. Then, after obtaining the system's state space model we are able to use the Kalman filter algorithm to estimate the future output of the system.

To study the dynamics of our system, we used MATLAB functions **etfe** and **spa** to firstly estimate the empirical transfer functions and then estimate the frequency response with fixed frequency resolution using spectral analysis. The continuous time-identified transfer function obtained is:

$$\frac{2.278\ s + 0.1711}{s^2 + 2.488\ s + 0.1695} \tag{52}$$

Using MATLAB we are able to acquire the Discrete-time identified state-space model:

$$\begin{aligned} x(t + Ts) &= A\ x(t) + B\ u(t) + K\ e(t) \\ y(t) &= C\ x(t) + D\ u(t) + e(t) \end{aligned} \tag{53}$$

with:

$$A = \begin{bmatrix} 0.8342 & -0.08908 \\ 0.0942 & -0.9716 \end{bmatrix}, B = \begin{bmatrix} 0.01966 \\ -0.03341 \end{bmatrix}, \quad C = [7.966 \quad 0.3005], D = 0, K = \begin{bmatrix} 0.006289 \\ -0.2834 \end{bmatrix} \tag{54}$$

Estimated using **N4SID** on time domain data. Fit to estimation data: 90.27% (prediction focus) with FPE: 0.4532 and MSE: 0.2292. **Figure 2** shows the

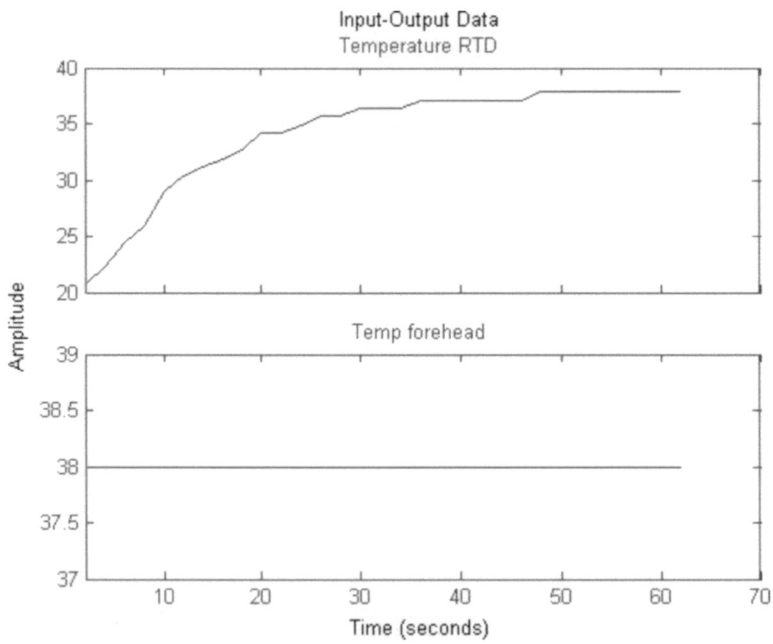

Figure 2.
System ID using MATLAB. Input output model for a step response defined problem.

Input–output model for which the input was set to a constant value of 38°C. The output, the step response, is that of the second order system as can be seen in the bode plot shown in **Figure 3**. **Figure 4** shows the evolution of the measured versus the modeled step responses.

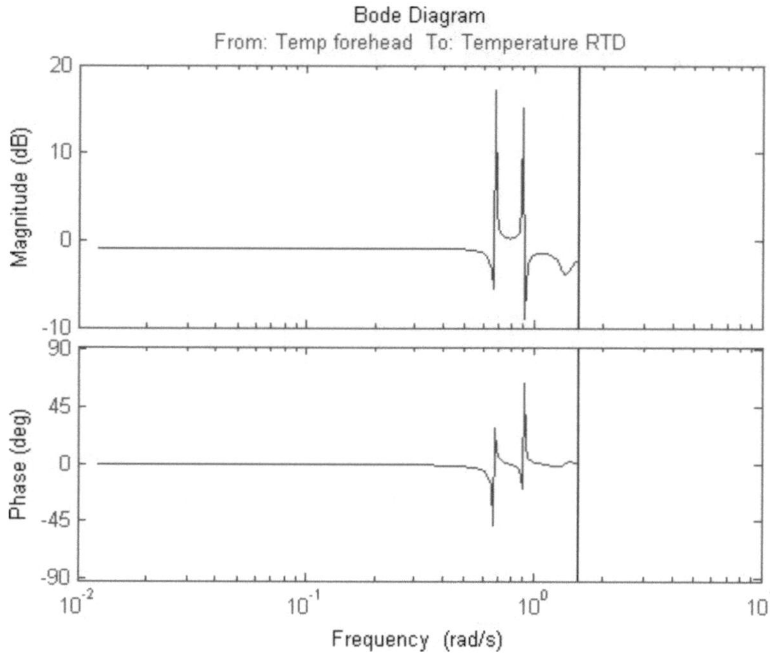

Figure 3.
Bode diagram indicating the system is a second order system as described by the system transfer function.

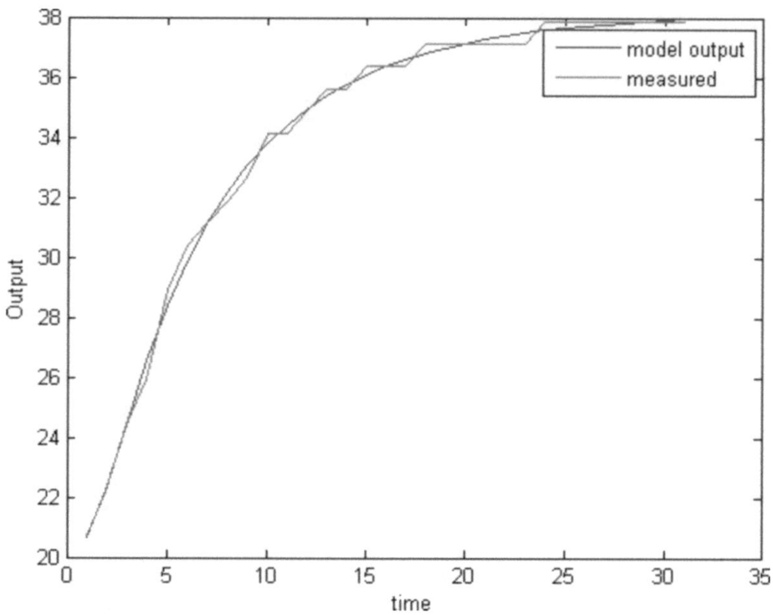

Figure 4.
Systems model and measured evolutions in time. Fit to estimation data: 90.27%.

4.2 Kalman filter

We modify the MATLAB example for the time-varying case found in [55] and we code our own function to solve the Discrete Algebraic Riccati Equation. MATLAB functions like predict or forecast were found useful to understand the problem at hand, however they were not used in the code we present here.

4.2.1 Time-varying Kalman filter using MATLAB

```
w(1:n) = sqrt(Q)*randn(n,1);
v(1:n) = sqrt(R)*randn(n,1);
systv = ss(A,B,C,0,Ts);
ytv(1:n) = lsim(systv,U(1:n) + w(1:n)).
yvtv(1:n) = ytv(1:n) + v(1:n);
Ptv(:,:) = B(:,:)*Q*B(:,:)';        % Initial error covariance.
x = zeros(order,1);      % Initial condition on the state.
order = 2;
yetv(1:n) = zeros(n,1);
ycov(1:n) = zeros(n,1);
for i = 1:n.
% Measurement update.
      Mn(:,:) = Ptv(:,:)*C(:,:)'/(C(:,:)*Ptv(:,:)*C(:,:)' + R);
      x = x + Mn(:,:)*(yvtv(i)-C(:,:)*x); % x[n|n].
      Ptv(:,:) = (eye(order)-Mn(:,:)*C(:,:))*Ptv(:,:);      % P[n|n].
      yetv(i) = C(:,:)*x;
      errcov(i) = C(:,:)*Ptv(:,:)*C(:,:)';
% Time update.
      x = A(:,:)*x + B(:,:)*U(i);       % x[n + 1|n].
      Ptv(:,:) = A(:,:)*Ptv(:,:)*A(:,:)' + B(:,:)*Q*B(:,:)'; P[n + 1|n].
end
%% DARE. We coded our own dare function [X,L,G] = sdare(A,B,Q,R).
[P_inf,L,M_inf] = sdare(atv,ctv',Q,R);
for i = 1:p
% Measurement update.
      x = x + M_inf*(yvtv(i)-ctv*x); % x[n|n].
yetv_inf(i) = ctv*x;
errcov_inf(i) = ctv*P_inf*ctv';
% Time update.
x = atv*x + btv*U(i); % x[n + 1|n].
P_inf = atv*P_inf*atv' + btv*Q*btv'; % P[n + 1|n].
end

function [SD] = sdare(A,B,Q,R).
At = transpose(A);
Bt = transpose(B);
S1 = size(A);
E = eye(S1);
Z = zeros(S1);
Ri = inv.(R);
S = B*Ri*Bt;
Pdare = [A Z; −Q E];
Ndare = [E S; Z At];
```

```
[AA,BB,L,Z] = qz(Pdare,Ndare);
[AAS1,BBS1,QS1,ZS1] = ordqz(AA,BB,L,Z,'udi');
O = ZS1(1:2,1:2);
P = ZS1(3:4,1:2);
H = inv.(O);
SD = P*H;
end
```

5. Simulations

Matlab was used to simulate the response of an RTD modelled as a second order system. In **Figure 5(A)** we show the plot of the true response y (cyan line) and the filtered response (red line). In **Figure 5(B)** the plot compares the measurement error with the estimation error. As can be seen in **Figure 5(C)** the time-varying filter also estimates the covariance errcov of the estimation error at each sample which shows when the filter reached steady state. As it can be seen, we have the possibility to predict the state after approximately 8 seconds. Also, we show the evolution of the estimated temperature response showing an error of -0.0948°C in the best of the cases and less than 1°C in the worst of the cases after 45 seconds.

6. Implementation

The unit step response depends on the roots of the characteristic equation. If both roots are real-valued, the second-order system behaves like a chain of two first-order systems, and the step response has two exponential components. If the roots are complex, the step response is a harmonic oscillation with an exponentially decaying amplitude [56]. In our case, the roots of the characteristic polynomial: $s^2 + 2.488\ s + 0.1695$ are -2.4179 and $-\ 0.0701$. Thus our system behaves like two first order systems in series.

The state description for an RC system is described above. From there, we know that the dynamics are dependent only on the RC constant. In addition, there is an amplificator in the system electronics that has a gain of 260. To solve for the RC constant of the system we use the least-squares method (Chi square minimization). The system has a solution of the form $y = e^{Bx}$ and we take n data points to form the vectors x_i and Y_i. The problem is to minimize the error function, $err = \sum_{i=1}^{n} \left(Y_i - Ae^{Bx_i}\right)^2$. The trick on the algorithm goes as follows:

$$Yn_i = ln\, Y_i = ln\left(Ae^{Bx_i}\right) = lnA + \ln\left(e^{Bx_i}\right) = C + Bx_i \tag{55}$$

Which is a linear equation. Using a linear fitting program:

$$B = \frac{n\sum x_i y_i - \sum x_i \sum y_i}{n\sum x_i^2 - \left(\sum x_i\right)^2} \text{ and } c = \frac{\sum x_i^2 \sum y_i - \sum x_i \sum x_i y_i}{n\sum x_i^2 - \left(\sum x_i\right)^2} \tag{56}$$

We obtain B and $A = \exp(c)$. We coded two Kalman filters with different calibration parameters (Q and R) as is written below. The KF was implemented in an ATMEGA328 microprocessor. Code for an Arduino was generated using C language which is available in the following section. Only the relevant portion is written.

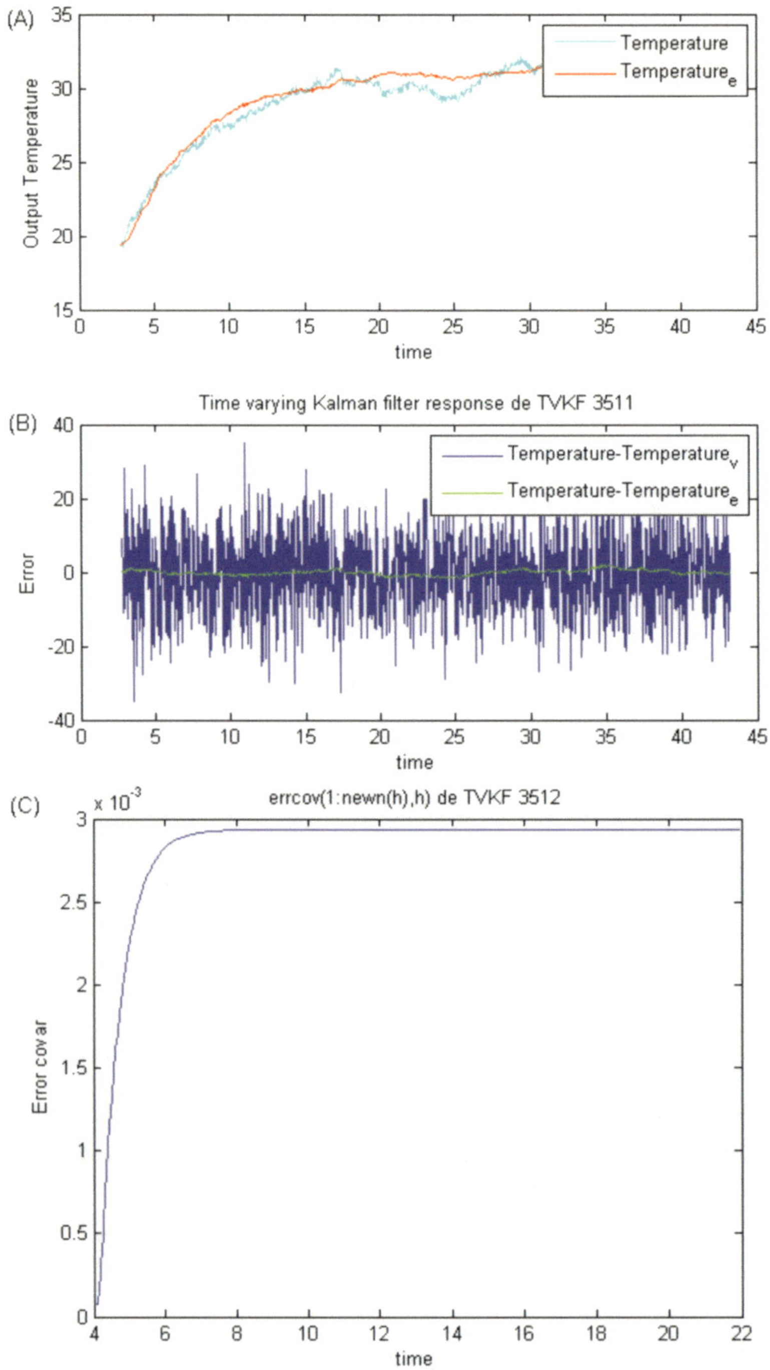

Figure 5.
(A) Evolution of the estimated temperature response showing an error of −0.0948°C in the best of the cases and less than 1°C in the worst of the cases. (B) Evolution of the measurement and estimation errors. (C) Evolution of the covariance of the error showing the possibility to predict the state after approximately 8 seconds.

6.1 Arduino code

```
readings[readIndex] = analogRead(inputPin); // read from the sensor.
total = total + readings[readIndex]; // add the reading to the total.
readIndex = readIndex +1; // advance to the next position in the array.
```

```
time_equis_readings[time_equis_readIndex] = time_equis_readIndex;
time_equis_readIndex = time_equis_readIndex +1;
if (readIndex > = numReadings) // if we're at the end of the array.
{
    for(i = 0;i < =numReadings-1;i++).
     {
     Y[i] = log(readings[i]);
     time1[i] = time_equis_readings[i];
     sumx = (sumx +time_equis_readings[i]);
     sumx2 = (sumx2 + time_equis_readings[i]*time_equis_readings[i]);
     sumy = (sumy +Y[i]);
     sumxy = (sumxy +time_equis_readings[i]*Y[i]);
     }
 den = (numReadings*sumx2-sumx*sumx);
 a = (sumx2*sumy -sumx*sumxy)/den;
 Bc = (n*sumxy-sumx*sumy)/den;
// State description.
 A = -Bc;B=Bc;C = 260;D = 0;
//wrap around to the beginning:
 readIndex = 0;time_equis_readIndex = 0;
 }
// KALMAN.
 errcov = C*P*C;
 for(i = 0;i < =numReadings-1;i++).
  {
    Mn = P*C/((C*P*C + R)); // initial estimate.
    X = X + Mn*(readings[i]-C*X); // update estimate Average_readings[i].
    P = (1-Mn*C)*P; // update covariance.
y_e[i] = C*X;
    errcov = C*P*C;
X = A*X + B*U; // project into k + 1.
    P = A*P*A + B*Q*B; // project into k + 1.
}
  timer0_millis = millis();
  // Solution to the Riccati equation.
  F = -Bc;H = 260;
  SQ = sqrt((H*H*Q*R) + (F*F*R*R));
SR = F * R;
P_inf = (SQ + SR)/(H*H);
M_inf = P_inf*C/(C*P_inf*C + R);
for(i = 0;i < =numReadings-1;i++).
  {
    // Measurement update.
    // M_inf;
    x_inf = x_inf + M_inf*(readings[i]-C*x_inf); // % x[n|n].
//P_inf; % P[n|n].
    y_e_inf = C*x_inf;
errcov_inf = C*P_inf*C;
    // Time update.
    x_inf = A*x_inf + B*U;
    P_inf = A*P_inf*A + B*Q*B;
  }
}
```

7. Results

The experiments were performed in a thermal bath giving step responses to the desired setup temperature. **Figure 6** depicts the upward and downward evolution of the temperature, the Kalman filter and the two predictors (using two different Q and R settings). As can be seen the predictors follow the Temperature of the sensor closely, especially for the upward way, while the Kalman filter lags behind.

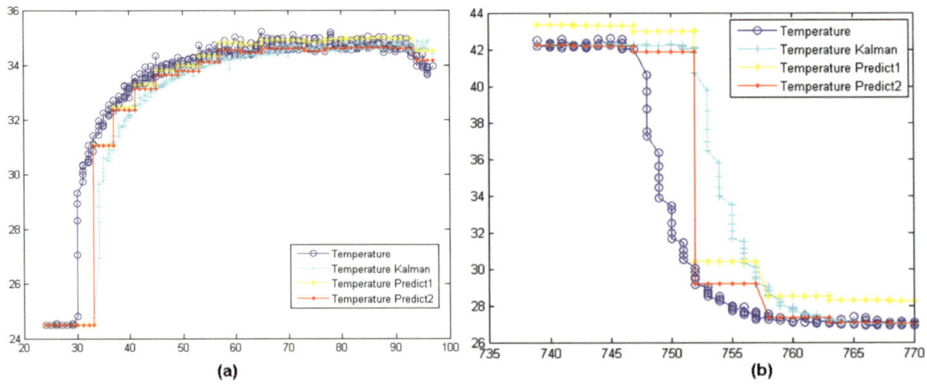

Figure 6.
Implemented system. Step response for the upwards and downwards evolution. Two different Kalman filters were used to predict (by solving the DARE equation) the evolution of the future state with different Q and R to calibrate the desired response. In blue the evolution of the RTD sensor analog input, in Yellow and red the two Kalman predictors and the Kalman estimation in cyan color.

8. Boundary layer

Sliding control [57] is an additional tool to predict the behavior of a second order system basically smoothing the system by boundary layers. The prediction of the system state trajectory is given using an uncertain model of the system. The subspace which represents the quantity of uncertainties in the prediction process, forces the estimate state trajectory to switching gain to converge the estimates to within a boundary of the real state values. To predict the state trajectory of our RLC system it's possible to switch its gain by the subspace represented by a first-order RC model. The estimated state trajectory is forced to keep a switch back and forth

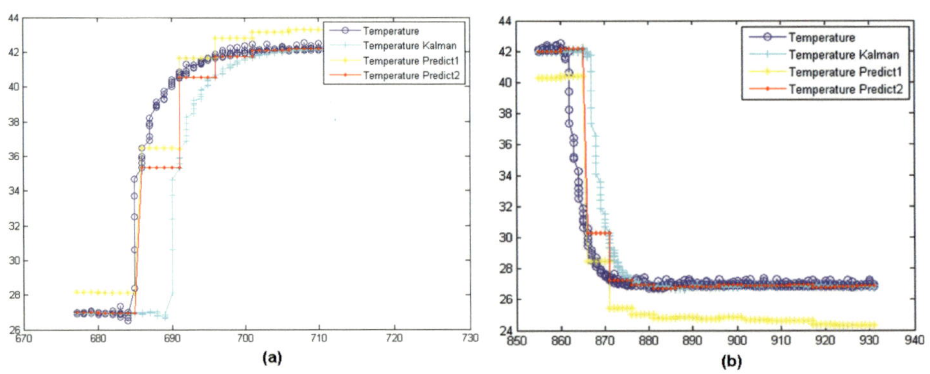

Figure 7.
Ascending and descending step responses of the Kalman filter and two Predictors which function in real time. In blue the RTD sensor response, in cyan the estimator response, in yellow and in red the two differently calibrated Kalman predictors.

within the boundary layer represented in our case by a RC model. By creating a boundary layer, the system is further constrained to have a solution existing in between two RC model solutions.

In **Figure 7** it can be clearly seen that the use of two estimators may help predict the behavior of the RTD in a much better way. The system needs to be calibrated first in order to have the two Kalman filters enveloping the required solution. As can be seen in the upward direction, both predictors (yellow and red) envelope the desired response (blue), that of the RTD sensor improving the response of the Kalman filter without boundaries (cyan). Unfortunately, this is not the case in the downward evolution. From the nonlinear control systems point of view these two evolutions demark a region where the RTD stands thus making possible to program a better estimator. It is left as an outlook to program a third estimator using this boundary layer in order to have a better predictor, especially for the downward evolution.

9. Conclusions

As it can be shown the implementation of the Kalman filter brings the opportunity to estimate the forecast in real time of a second order system using first, MATLAB and second that of two first order systems using a simple RC system coded in C-language for a microprocessor. It has been shown that the program is able to predict the evolution of temperature for a RTD system. Even if the system is implemented using a first order system we can find evolving solutions for our estimation and prediction to be good enough. We predict the state after approximately 8 seconds showing an error of -0.0948°C in the best of the cases. In addition, a boundary layer may be programmed using two first order Kalman predictors which may be tuned by setting Q and R properly. We believe this is the first report on the use of a Kalman filter to predict the evolution of temperature from a RTD.

Author details

Erick Ulin-Avila* and Juan Ponce-Hernandez
Center for Engineering and Industrial Development, Querétaro, México

*Address all correspondence to: ulinavilaerick@gmail.com

References

[1] Xincun, Y.; Yongzhong, O.; Fuping, S.; Hui, F. Kalman Filter Applied in Underwater Integrated Navigation System Underwater Integrated Navigation System Profile Kalman Filter 4 Kalman Rllter Integrated Navigation System of Underwater. Geod. Geodyn. **2013**, *4* (1), 46–50. https://doi.org/10.3724/SP.J.1246.2013.01046.

[2] Popov, I.; Koschorrek, P.; Haghani, A.; Jeinsch, T. Adaptive Kalman Filtering for Dynamic Positioning of Marine Vessels. IFAC-PapersOnLine **2017**, *50* (1), 1121–1126. https://doi.org/10.1016/j.ifacol.2017.08.394.

[3] Zhao, Y. Performance Evaluation of Cubature Kalman Filter in a GPS/IMU Tightly-Coupled Navigation System. Signal Processing **2016**, *119*, 67–79. https://doi.org/10.1016/j.sigpro.2015.07.014.

[4] Allotta, B.; Caiti, A.; Costanzi, R.; Fanelli, F.; Fenucci, D.; Meli, E.; Ridolfi, A. A New AUV Navigation System Exploiting Unscented Kalman Filter. Ocean Eng. **2016**, *113*, 121–132. https://doi.org/10.1016/j.oceaneng.2015.12.058.

[5] Khan, N.; Bacha, S. A.; Khan, S. A. Improvement of Compensated Closed-Loop Kalman Filtering Using Autoregressive Moving Average Model. Measurement **2019**, *134*, 266–279. https://doi.org/10.1016/j.measurement.2018.10.063.

[6] Yuan, J.; Wang, Y.; Ji, Z. A Differentially Private Square Root Unscented Kalman Filter for Protecting Process Parameters in ICPSs. ISA Trans. **2020**, *104*, 44–52. https://doi.org/10.1016/j.isatra.2019.12.010.

[7] Hamiche, K.; Abouaïssa, H.; Goncalves, G.; Hsu, T. A Robust and Easy Approach for Demand Forecasting in Supply Chains. IFAC-PapersOnLine **2018**, *51* (11), 1732–1737. https://doi.org/10.1016/j.ifacol.2018.08.206.

[8] Baradaran Khalkhali, M.; Vahedian, A.; Sadoghi Yazdi, H. Vehicle Tracking with Kalman Filter Using Online Situation Assessment. Rob. Auton. Syst. **2020**, *131*, 103596. https://doi.org/10.1016/j.robot.2020.103596.

[9] Farahi, F.; Yazdi, H. S. Probabilistic Kalman Filter for Moving Object Tracking. Signal Process. Image Commun. **2020**, *82* (December 2019), 115751. https://doi.org/10.1016/j.image.2019.115751.

[10] Piovoso, M.; Laplante, P. A. Kalman Filter Recipes for Real-Time Image Processing. Real-Time Imaging **2003**, *9* (6), 433–439. https://doi.org/10.1016/j.rti.2003.09.005.

[11] Hamuda, E.; Mc Ginley, B.; Glavin, M.; Jones, E. Improved Image Processing-Based Crop Detection Using Kalman Filtering and the Hungarian Algorithm. Comput. Electron. Agric. **2018**, *148* (February), 37–44. https://doi.org/10.1016/j.compag.2018.02.027.

[12] Wang, L.; Loffeld, O.; Ma, K.; Qian, Y. Sparse ISAR Imaging Using a Greedy Kalman Filtering Approach. Signal Processing **2017**, *138*, 1–10. https://doi.org/10.1016/j.sigpro.2017.03.002.

[13] Predictor, K. On-Line Temperature Estimation for Noisy Thermal Sensors Using a Smoothing Filter-Based. **2018**. https://doi.org/10.3390/s18020433.

[14] Shrivastava, P.; Soon, T. K.; Idris, M. Y. I. Bin; Mekhilef, S. Overview of Model-Based Online State-of-Charge Estimation Using Kalman Filter Family for Lithium-Ion Batteries. *Renew. Sustain. Energy Rev.* **2019**, *113* (December 2018), 109233. https://doi.org/10.1016/j.rser.2019.06.040.

[15] Linghu, J.; Kang, L.; Liu, M.; Luo, X.; Feng, Y.; Lu, C. Estimation for State-of-Charge of Lithium-Ion Battery Based on an Adaptive High-Degree Cubature Kalman Filter. *Energy* **2019**, *189* (xxxx), 116204. https://doi.org/10.1016/j.energy.2019.116204.

[16] Zhang, S.; Guo, X.; Zhang, X. An Improved Adaptive Unscented Kalman Filtering for State of Charge Online Estimation of Lithium-Ion Battery. J. Energy Storage **2020**, *32* (September), 101980. https://doi.org/10.1016/j.est.2020.101980.

[17] Sassi, H. Ben; Errahimi, F.; Es-sbai, N. State of Charge Estimation by Multi-Innovation Unscented Kalman Filter for Vehicular Applications. *J. Energy Storage* **2020**, *32* (October), 101978. https://doi.org/10.1016/j.est.2020.101978.

[18] Wang, H.; Lei, T.; Rong, Y.; Shao, W.; Huang, Y. Arc Length Stable Method of GTAW Based on Adaptive Kalman Filter. *J. Manuf. Process.* **2020**, No. December 2019, 0–1. https://doi.org/10.1016/j.jmapro.2020.01.029.

[19] Holtz, J. Sensorless Control of Induction Machines – with or without Signal Injection ? **2019**, No. July. https://doi.org/10.1109/TIE.2005.862324.

[20] Ameid, T.; Menacer, A.; Talhaoui, H.; Harzelli, I. Rotor Resistance Estimation Using Extended Kalman Filter and Spectral Analysis for Rotor Bar Fault Diagnosis of Sensorless Vector Control Induction Motor. Meas. J. Int. Meas. Confed. **2017**, *111*, 243–259. https://doi.org/10.1016/j.measurement.2017.07.039.

[21] Chen, Z.; Wang, L.; Liu, X. *Sensorless Direct Torque Control of PMSM Using Unsected Kalman Filter*; IFAC, 2011; Vol. 44. https://doi.org/10.3182/20110828-6-IT-1002.02515.

[22] Nilsson, T.; Soja, B.; Karbon, M.; Heinkelmann, R.; Schuh, H. Application of Kalman Filtering in VLBI Data Analysis. Earth, Planets Sp. **2015**. https://doi.org/10.1186/s40623-015-0307-y.

[23] Karbon, M.; Soja, B.; Nilsson, T.; Deng, Z.; Heinkelmann, R.; Schuh, H. Earth Orientation Parameters from VLBI Determined with a Kalman Filter. Geod. Geodyn. **2017**, *8* (6), 396–407. https://doi.org/10.1016/j.geog.2017.05.006.

[24] Teh, L. A. and J. Kalman Filter for Reducing Total Harmonics Distortion in Stand-Alone PV System. *2020 Glob. Congr. Electr. Eng. (GC- ElecEng)* **2020**, 81–87. https://doi.org/10.23919/GC-ElecEng48342.2020.9286275.Abstract.

[25] Docimo, D. J.; Ghanaatpishe, M.; Mamun, A. Extended Kalman Filtering to Estimate Temperature and Irradiation for Maximum Power Point Tracking of a Photovoltaic Module. Energy **2017**, *120*, 47–57. https://doi.org/10.1016/j.energy.2016.12.089.

[26] Monteiro, R. V. A.; Guimarães, G. C.; Moura, F. A. M.; Albertini, M. R. M. C.; Albertini, M. K. Estimating Photovoltaic Power Generation: Performance Analysis of Artificial Neural Networks, Support Vector Machine and Kalman Filter. Electr. Power Syst. Res. **2017**, *143*, 643–656. https://doi.org/10.1016/j.epsr.2016.10.050.

[27] Madhukar, P. S. S. M. Overview. **2020**, No. Icosec, 1268–1272.

[28] Belkhatir, Z.; Mechhoud, S.; Laleg-Kirati, T. M. Kalman Filter Based Estimation Algorithm for the Characterization of the Spatiotemporal Hemodynamic Response in the Brain. Control Eng. Pract. **2019**, *89* (May), 180–189. https://doi.org/10.1016/j.conengprac.2019.05.017.

[29] Alguliyev, R.; Imamverdiyev, Y.; Sukhostat, L. Cyber-Physical Systems

and Their Security Issues. Comput. Ind. **2018**, *100* (July 2017), 212–223. https://doi.org/10.1016/j.compind.2018.04.017.

[30] Wang, J.; Luo, J.; Liu, X.; Li, Y.; Liu, S.; Zhu, R. Improved Kalman Filter Based Differentially Private Streaming Data Release in Cognitive Computing. Futur. Gener. Comput. Syst. **2019**, *98*, 541–549. https://doi.org/10.1016/j.future.2019.03.050.

[31] Zhang, Y.; Wang, R.; Li, S.; Qi, S. Temperature Sensor Denoising Algorithm Based on Curve Fitting and Compound Kalman Filtering. Sensors (Switzerland) **2020**, *20* (7), 1–13. https://doi.org/10.3390/s20071959.

[32] Mouzinho, L. F.; Fonsecaneto, J. V.; Luciano, B. A.; Freire, R. C. S. Indirect Measurement of the Temperature via Kalman Filter. *18th IMEKO World Congr. 2006 Metrol. a* Sustain. Dev. **2006**, *1*, 818–823.

[33] Ma, Y.; Cui, Y.; Mou, H.; Gao, J.; Chen, H. Core Temperature Estimation of Lithium-Ion Battery for EVs Using Kalman Filter. Appl. Therm. Eng. **2020**, *168* (April 2019), 114816. https://doi.org/10.1016/j.applthermaleng.2019.114816.

[34] Eleffendi, M. A.; Johnson, C. M. Application of Kalman Filter to Estimate Junction Temperature in IGBT Power Modules. IEEE Trans. Power Electron. **2016**, *31* (2), 1576–1587. https://doi.org/10.1109/TPEL.2015.2418711.

[35] Wiener, N. *The Extrapolation, Interpolation and Smoothing of Stationary Time Series with Engineering Applications*; John Wiley & Sons: New York, 1949.

[36] Kalman, R. E. A New Approach to Linear Filtering and Prediction Problems. J. Fluids Eng. Trans. ASME **1960**, *82* (1), 35–45. https://doi.org/10.1115/1.3662552.

[37] Tewari, A. Modern Control Design With MATLAB and SIMULINK. **2002**, 518.

[38] Lacey, T. Tutorial : The Kalman Filter. 133–140.

[39] Grewal, M. S.; Andrews, A. P. *Kalman Filtering: Theory and Practice Using MATLAB®: Third Edition*; 2008. https://doi.org/10.1002/9780470377819.

[40] Wang, L. *Model Predictive Control System Design and Implementation Using MATLAB*; 2009; Vol. 53.

[41] Lu, L. Z.; Lin, W. W. An Iterative Algorithm for the Solution of the Discrete-Time Algebraic Riccati Equation. Linear Algebra Appl. **1993**, *188–189* (C), 465–488. https://doi.org/10.1016/0024-3795(93)90476-5.

[42] Laudadio, T.; Mastronardi, N.; Van Dooren, P. The Generalized Schur Algorithm and Some Applications. Axioms **2018**, *7* (4), 1–18. https://doi.org/10.3390/axioms7040081.

[43] DATTA, B. N. Numerical Solutions and Conditioning of Algebraic Riccati Equations. *Numer. Methods Linear Control Syst.* **2004**, 519–599. https://doi.org/10.1016/b978-012203590-6/50017-3.

[44] Chauhan, J.; Neelakantan, U. An Experimental Approach for Precise Temperature Measurement Using Platinum RTD PT1000. Int. Conf. Electr. Electron. Optim. Tech. ICEEOT *2016* **2016**, 3213–3215. https://doi.org/10.1109/ICEEOT.2016.7755297.

[45] Trung, T. Q.; Ramasundaram, S.; Hwang, B. U.; Lee, N. E. An All-Elastomeric Transparent and Stretchable Temperature Sensor for Body-Attachable Wearable Electronics. Adv. Mater. **2016**, *28* (3), 502–509. https://doi.org/10.1002/adma.201504441.

[46] Chen, Y.; Lu, B.; Chen, Y.; Feng, X. Breathable and Stretchable Temperature Sensors Inspired by Skin. Sci. Rep. **2015**, *5*, 1–11. https://doi.org/10.1038/srep11505.

[47] Wang, Z.; Gao, W.; Zhang, Q.; Zheng, K.; Xu, J.; Xu, W.; Shang, E.; Jiang, J.; Zhang, J.; Liu, Y. 3D-Printed Graphene/Polydimethylsiloxane Composites for Stretchable and Strain-Insensitive Temperature Sensors. ACS Appl. Mater. Interfaces **2019**, *11* (1), 1344–1352. https://doi.org/10.1021/acsami.8b16139.

[48] Kim, J.; Kim, J.; Shin, Y.; Yoon, Y. A Study on the Fabrication of an RTD (Resistance Temperature Detector) by Using Pt Thin Film. Korean J. Chem. Eng. **2001**, *18* (1), 61–66. https://doi.org/10.1007/BF02707199.

[49] Noh, J.; Park, S.; Boo, H.; Kim, H. C.; Chung, T. D. Nanoporous Platinum Solid-State Reference Electrode with Layer-by-Layer Polyelectrolyte Junction for PH Sensing Chip. Lab Chip **2011**, *11* (4), 664–671. https://doi.org/10.1039/c0lc00293c.

[50] Hassan, A. S.; Juliet, V.; Joshua Amrith Raj, C. MEMS Based Humidity Sensor with Integration of Temperature Sensor. *Mater.* Today Proc. **2018**, *5* (4), 10728–10737. https://doi.org/10.1016/j.matpr.2017.12.356.

[51] Imran, M.; Bhattacharyya, A. Thermal Response of an On-Chip Assembly of RTD Heaters, Sputtered Sample and Microthermocouples. Sensors Actuators, A Phys. **2005**, *121* (2), 306–320. https://doi.org/10.1016/j.sna.2005.02.019.

[52] Kang, L.; Shi, Y.; Zhang, J.; Huang, C.; Zhang, N.; He, Y.; Li, W.; Wang, C.; Wu, X.; Zhou, X. A Flexible Resistive Temperature Detector (RTD) Based on in-Situ Growth of Patterned Ag Film on Polyimide without Lithography. Microelectron. Eng. **2019**, *216* (July), 111052. https://doi.org/10.1016/j.mee.2019.111052.

[53] Cui, J.; Liu, H.; Li, X.; Jiang, S.; Zhang, B.; Song, Y.; Zhang, W. Fabrication and Characterization of Nickel Thin Film as Resistance Temperature Detector. Vacuum **2020**, *176*, 109288. https://doi.org/10.1016/j.vacuum.2020.109288.

[54] Lee, Y.; Cheng, S.; Fang, W. MONOLITHIC INTEGRATED CMOS-MEMS FLUORESCENCE QUENCHING GAS SENSOR AND RESISTIVE TEMPERATURE DETECTOR (RTD) FOR TEMPERATURE COMPENSATION. *2019 20th Int. Conf. Solid-State Sensors, Actuators Microsystems Eurosensors XXXIII (TRANSDUCERS EUROSENSORS XXXIII)* **2019**, No. June, 1293–1296.

[55] Mathworks. Kalman filtering.

[56] Haidekker, M. A. Solving Differential Equations in the Laplace Domain. Linear Feed. Control. **2013**, 27–56. https://doi.org/10.1016/b978-0-12-405875-0.00003-6.

[57] Gadsden, S. A.; Eng, B. M. Smooth Variable Structure Filtering: Theory and Applications. *Thesis* **2011**.

Chapter 4

A Constant Gain Kalman Filter for Wireless Sensor Network and Maneuvering Target Tracking

Peeyush Awasthi, Ashwin Yadav, Naren Naik and Mudambi Ramaswamy Ananthasayanam

Abstract

One of the well-known approaches to target tracking is the Kalman filter. The problem of applying the Kalman Filter in practice is that in the presence of unknown noise statistics, accurate results cannot be obtained. Hence the tuning of the noise covariances is of paramount importance in order to employ the filter. The difficulty involved with the tuning attracts the applicability of the concept of Constant Gain Kalman Filter (CGKF). It has been generally observed that after an initial transient the Kalman Filter gain and the State Error Covariance P settles down to steady state values. This encourages one to consider working directly with steady state or constant Kalman gain, rather than with error covariances in order to obtain efficient tracking. Since there are no covariances in CGKF, only the state equations need to be propagated and updated at a measurement, thus enormously reducing the computational load. The current work first applies the CGKF concept to heterogeneous sensor based wireless sensor network (WSN) target tracking problem. The paper considers the Standard EKF and CGKF for tracking various manoeuvring targets using nonlinear state and measurement models. Based on the numerical studies it is clearly seen that the CGKF out performs the Standard EKF. To the best of our knowledge, such a comprehensive study of the CGKF has not been carried out in its application to diverse target tracking scenarios and data fusion aspects.

Keywords: Constant Gain Kalman Filter, INS, GPS, Wireless Sensor Network, Tracking

1. Introduction

The Kalman Filter (KF) is one of the most fundamental and widely used estimation schemes in tracking application. While the KF formalism is very powerful we need to keep in mind that the solution scheme can be considered to be formal and a fundamental prerequisite for accurate results is the a prioiri knowledge of the initial state (X_0), initial state noise covariance (P_0), system noise covariance (Q), measurement noise covariance (R). Good values of X_0, P_0, Q and R are imperative for the filter to perform optimally. Tuning of the KF is defined as the process to obtain precise values of P_0, Q, and R. A detailed review of filter tuning has been given by Ananthasayanam et al. in [1]. A main theme of filter covariance tuning schemes is the notion of the innovation sequence being white and Gaussian for filter optimality.

One class of schemes obtains the unknown covariances that maximize the likelihood [2–5]. Another important class of filter covariance tuning schemes is the covariance matching methodology [6–8]. The idea of an innovations based cost function being minimized by the optimal covariances is also used in [7, 8] to tune process system noise (Q) and state error covariances (P) respectively.

An alternate approach to tuning is via the direct setting of the Kalman gain as carried out in the work of Ananthasayanam et al. [9] and Ashwin et al. [10]. It is often observed that the Kalman gain converges to a steady state value which coincides with the convergence of the state error covariance P. The premise for working with the steady state or constant gain is well explained in the thesis work by Bohn [11]. The work of Anil Kumar et al. [9], optimizes the innovations likelihood cost function [5] for the (constant) Kalman gain in a space craft reentry problem. This CGKF approach works directly with the Kalman gain and does not utilize any knowledge of the filter covariances.

Our present work is about the application and sensitivity study of the CGKF traget tracking scheme in sensor networks scenarios and maneuvering target tracking. We look at target tracking problems in wireless sensor networks using passive infrared (PIR), acoustic and seismic sensors in stand alone (SA) and data fusion (DF) modes as given by Raol [12] for the discrete white noise acceleration (DWNA) traget motion model. We further demonstrate the capability of the CGKF to track maneuvering targets [13, 14] from acquired range and direction data for a class of coordinated turn (CT) maneuver models. The CGKF with linear measurement model was validated in [10, 15]. The present study applies the CGKF to a non linear measurement models and further demonstrates its robustness through sensitivity studies. The results obtained with respect to homogeneous and heterogeneous data fusion further demonstrate the range of applicability of the CGKF. These extensive tracking and sensitivity studies for a wide range of state and measurement models are to the best of our knowledge, unique to this paper and provide the reader with a comprehensive reference. These results also provide a firm base for application of the CGKF concept to other areas. In the sequel, Section 2 describes the CGKF concept. Section 3 introduces the various tracking scenarios based on PIR, acoustic and seismic measuerement models in SA and DF modes. In addition maneuvering targets based on CT models are discussed, since these have the potential to demonstrate the flexibility of the CGKF. Section 4 details the tracking and sensitivity studies on the above mentioned models, and Section 5 gives the conclusion of the present work.

2. Constant gain Kalman filtering

The KF algorithm [16] is based on the least squares principle with recursive time updates. It is a fact that optimal filter performance needs apriori knowledge of the filter statistics in terms of the state-error, system and measurement noise covariances (P, Q and R respectively). A central theme in the optimality of the KF is the requirement of the innovations being white at convergence [17, 18]. Mehra [17] shows that the settling of the filter gain value to a steady state value coincides with the state error covariance also similarly settling (**Figure 1**).

The observation that the gain (reflecting P, Q, R) reaches a steady state, prompts us to consider working directly with the steady state gain rather than the tuning dependant P, Q, R matrices to determine the gain. The way we accomplish this is via an innovations cost function minimization approach. We use the whiteness of the innovations at KF convergnce in order to construct the likelihood based function of the innovations sequence [5].

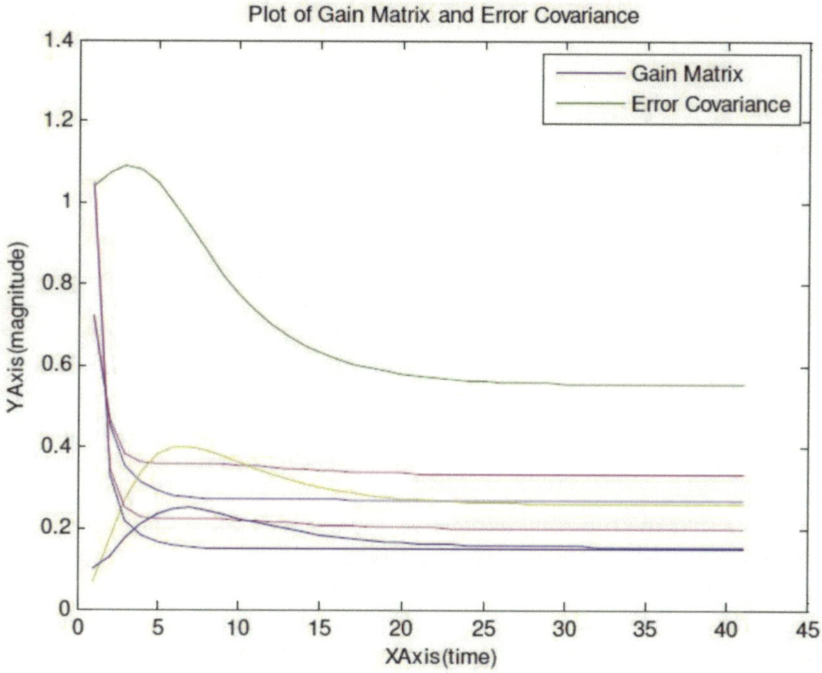

Figure 1.
Gain K *vs. error covariance matrix* P.

$$J(K,\mathcal{R}) = \frac{1}{N}\sum_{t=1}^{N}\left(v_t^T\mathcal{R}v_t + log\left(|\mathcal{R}|\right)\right) \tag{1}$$

where v_t represents the innovations, $\mathcal{R}$ represents the innovations covariance, $|.|$ represents the determinant and N is the number of measurement time steps. We obtain the steady state gain K^* and innovation covariance $\mathcal{R}^*$ by solving the following optimization problem

$$(K^*,\mathcal{R}^*) = argmin_{K,\mathcal{R}}J(K,\mathcal{R}) \tag{2}$$

The following is the estimation scheme based on a predict and update mode.

2.1 The estimation scheme

The generic KF updates are

$$\hat{x}_t = \overline{x}_t + K_t v_t \tag{3}$$

where the the innovations sequence is $v_t = y_t - C\overline{x}_t$. C is the measurement matrix, $\overline{x}_t$ is the predicted state matrix and $\hat{x}_t$ is the filtered state matrix. The standard KF computes the gain matrix K_t using P, Q and R while we proceed to estimate this constant gain K^* for the CGKF, by solving the optimization problem described by (2) above. The optimization problem can be solved using local gradient based methods (such as Newton type schemes) [19] or global schemes such as Genetic Algorithm (GA) [20] applied to problems as in [9]. As the filter tracks the target, the gain K is seen to stabilize to a value given by the solution of the above problem. Once we have computed the optimal filter gain K_t (denoted henceforth by K^*, representative of the constant gain) for the CGKF, the KF recursions become.

Predict

$$\overline{x}_{t+1} = A\hat{x}_{t+1} + u_{t+1} \tag{4}$$

Update

$$\hat{x}_{t+1} = \overline{x}_{t+1} + K^* \left(y_{t+1} - C\overline{x}_{t+1}\right) \tag{5}$$

Thus it is evident that once the optimal gain K^* is computed using GA to solve the optimization problem, the filter algorithm reduces to a simple predict and update model given by (4) and (5) above. This is obviously more compact compared to the standard KF which is implemented in five steps involving computation and propagation of the State Error Covariance P using Q and R. The advantage is speed of operation because we circumvent the tedious calculations of the costly covariance matrices P, Q, R and instead work directly with the optimal gain for the set of measurements.

We observe that the typically expensive covariance time update step is not needed in the constant gain approach.The CGKF is found to work quite well even with state models moderately different from that for which the gains are computed [25], suggesting a robustness of the gains calculated (Refer **Tables 6** and 7). It is to be noted that the present problem is a non linear problem, in so far as the measurement model is concerned so that the filter used is the CGKF. This is one unique advantage of the CGKF over the standard KF/EKF wherein the EKF requires linearization of the measurement model via use of the Jacobian H. The reconstruction in CGKF case employing the GA as the optimization tool, does not rely on the Jacobian in computation of the optimum Constant Filter Gain K^*.

3. Sensor models and modes

The focus of our study is the application of the CGKF to a variety of 2D sensor models such as those in unattended ground sensor (UGS) and Intelligence, Surveillance and Reconnaissance (ISR) systems. Sensors such as passive infrared (PIR) [21], acoustic, seismic [22] and radar have been studied. The sensor system might consist of single or multiple data inputs as required in different scenarios. They may consist of single type of sensor or multiple type of sensor nodes, as required in situations. Homogeneous and heterogeneous DF aspects of certain combination of sensors will be analyzed. We outline the regular and CGKF schemes and their application to the above mentioned systems.

3.1 State variable models in stand alone mode

Non Maneuvering or Discrete White Noise Acceleration (DWNA) Model.

The state model for 2D target is comprised of x and y direction displacement and their corresponding velocities wherein the state vector is represented as $X_t = \left[x_t, y_t, \dot{x}_t, \dot{y}_t\right]^t$ with x_t, y_t representing X,Y coordinates respectively of the target and $\dot{x}_t, \dot{y}_t$ representing velocities in X, Y directions.

State Equation.

The state equation for the DWNA is

$$X_{t+1} = AX_t + Bw_t \tag{6}$$

where

$$A = \begin{pmatrix} 1 & 0 & T_s & 0 \\ 0 & 1 & 0 & T_s \\ 0 & 0 & 1 & 0 \\ 0 & 0 & 0 & 1 \end{pmatrix}, B = \begin{pmatrix} 0.5T_s^2 & 0 \\ 0 & 0.5T_s^2 \\ T_s & 0 \\ 0 & T_s \end{pmatrix} \quad (7)$$

with A and B being the state - transition and acceleration matrices respectively, w_t being an uncorrelated Gaussian process.

Measurement equation: The measurement at time t of nth sensor $g_t^{(n)}$

$$g_t^{(n)} = h^{(n)}(X_t, t) + v_t^{(n)} \quad (8)$$

where $h^{(n)}(X_t, t)$ is typically a nonlinear function of the states $v^{(n)}$ is the corresponding measurement noise (assumed to be white Gaussian) of nth sensor. The measurement equations for the respective sensors are given below.

Sensor Measurement model for PIR sensor [21]

$$g_t^{(n)} = log \frac{\dot{x}_t^2 + \dot{y}_t^2}{\left(x_t - r_x^{(n)}\right)^2 + \left(y_t - r_y^{(n)}\right)^2} + v_t^{(n)} \quad (9)$$

Sensor Measurement model for Acoustic sensor [22]

$$g_t^{(n)} = tan^{-1}\left(y_t - r_y^{(n)} / x_t - r_x^{(n)}\right) + v_t^{(n)} \quad (10)$$

Sensor Measurement model for Seismic sensor [22]

$$g_t^{(n)} = \begin{pmatrix} \left(\left(x_t - r_x^{(n)}\right)^2 + \left(y_t - r_y^{(n)}\right)^2\right)^{0.5} + v_t^{(n)} \\ tan^{-1}\left(y_t - r_y^{(n)} / x_t - r_x^{(n)}\right) + v_t^{(n)} \end{pmatrix} \quad (11)$$

where $r^{(n)} = \left[r_x^{(n)}, r_y^{(n)}\right]$ is the position of the nth sensor in network.

Estimation Scheme.

We now outline the estimation scheme by an EKF as well as a CGKF. The EKF has the following steps. For t = 0,1,2......

Prediction

$$\overline{X}_{t+1} = A\hat{X}_t \quad (12)$$

$$\overline{P}_{t+1} = A\hat{P}_t A' + Q \quad (13)$$

where $\overline{X}_{t+1}$ is the predicted estimate based on the filtered estimate $\hat{X}_t$, where $\overline{P}_{t+1}$ and $\hat{P}_t$ being the state error covariances corresponding to the predicted $\overline{X}_{t+1}$ and filtered $\hat{X}_t$ estimates respectively. and $Q = BE(ww')B'$.

Update/Correction: The update of the states and covariances as per the EKF scheme are

$$\hat{X}_{t+1} = \overline{X}_{t+1} + K_{t+1}\left(g_{t+1} - h\left(\overline{X}_{t+1}, t\right)\right) \quad (14)$$

where

$$K_{t+1} = \overline{P}_{t+1}H_{t+1}'\left(H_{t+1}\overline{P}_{t+1}H_{t+1}\prime + R\right)^{-1} \tag{15}$$

where H_{t+1} is the Jacobian corresponding to $h(.)$ at time $t+1$ and $R = E(vv')$.

$$\hat{P}_{t+1} = (I - K_{t+1}H_{t+1})\overline{P}_{t+1} \tag{16}$$

Table 1 gives measurement Jacobians for all three sensors. In the table $\overline{d}_t = \left(\overline{x}_t - r_x^{(n)}\right)^{\overline{2}} + \left(\overline{y}_t - r_y^{(n)}\right)^2$ is used for sake of brevity of space.

The CGKF on the other hand has the following two steps.

Prediction

$$\overline{X}_{t+1} = A\hat{X}_{t+1} \tag{17}$$

Update/Correction.

Once the optimized Kalman gain K has been calculated via equations - 1,2 The following Eq. (17) updates the state parameters.

$$\hat{X}_{t+1} = \overline{X}_{t+1} + K\left(g_{t+1} - h\left(\overline{X}_t, t\right)\right) \tag{18}$$

In our work the optimized value of K is calculated via the application of the genetic algorithm to the innovation cost function Eqs. (1) and (2).

3.2 Homogeneous data fusion

In homogeneous fusion the fusion is based on the data from multiple sensors of similar type, at every time instant. Here in this section we have used mainly the centralized approach to DF in respect of the KF. The data obtained from various nodes (similar type of sensors) is combined together then applied to EKF and CGKF for tracking the target. This approach has been used as measurement fusion [12] approach in WSN of UGSs.

Measurement fusion techniques combine the raw measurements of the target obtained from the Individual Sensor Node (ISN) at the Cluster Head Node (CHN) Level. The ISN is a tier 1 node while the CHN is a tier 2 node which is capable of running a complex fusion algorithm based on KF framework. So ISNs are considered to have minimal computation capability compared to the CHNs. The two approaches which have been implemented in our work with respect to the CGKF under the

Sensors Type	Measurement Equation	Jacobian H
PIR	$h = log\,\frac{\overline{\dot{x}}_t^2 + \overline{\dot{y}}_t^2}{\left(\overline{x}_t - r_x^{(n)}\right)^2 + \left(\overline{y}_t - r_y^{(n)}\right)^2}.$	$H = \left[\frac{-2\left(\overline{x}_t - r_x^{(n)}\right)}{\overline{d}^2{}_t}, \frac{-2\left(\overline{y}_t - r_y^{(n)}\right)}{\overline{d}_t{}^2}, \frac{2\overline{\dot{x}}_t}{\overline{\dot{x}}_t^2 + \overline{\dot{y}}_t^2}, \frac{2\overline{\dot{y}}_t}{\overline{\dot{x}}_t^2 + \overline{\dot{y}}_t^2}\right]$
ACO-USTIC	$h = tan^{-1}\left(y_k - r_y^{(n)}/x_k - r_x^{(n)}\right)$	$H = \left[\frac{\left(r_y^{(n)} - \overline{y}_t\right)}{\overline{d}t}, \frac{\overline{x}_t - r_x^{(n)}}{\overline{d}_t}, 0, 0\right]$
SEISMIC	$h = \begin{pmatrix} \left(\left(x_t - r_x^{(n)}\right)^2 + \left(y_t - r_y^{(n)}\right)^2\right)^{0.5} + v_t^{(n)} \\ tan^{-1}\left(y_t - r_y^{(n)}/x_t - r_x^{(n)}\right) + v_t^{(n)} \end{pmatrix}$	$H = \left[\frac{\overline{x}_t - r_x^{(n)}}{\overline{d}_t^{0.5}}, \frac{\overline{y}_t - r_y^{(n)}}{\overline{d}_t^{0.5}}, 0, 0; \frac{\left(r_y^{(n)} - \overline{y}_t\right)}{\overline{d}_t}, \frac{\left(r_x^{(n)} - \overline{x}_t\right)}{\overline{d}_t}, 0, 0\right]$

Table 1.
Jacobians for sensors measurement models.

homogeneous DF are Maximal Kalman filter (MKF) [12] and Weighted fusion (WF) [12] approaches. The state model is the DWNA model of the previous sub section.

3.2.1 Maximal Kalman filter (MKF) method

This method is based on fusing all measurements of the ISN by incorporating them in a fused measurement vector and the corresponding measurement noise covariance and measurement matrices as described below

$$g_t^f = [g_t^1, g_t^2, \dots \dots g_t^m] \tag{19}$$

$$H_t^f = [H_t^1, H_t^2, \dots \dots H_t^m] \tag{20}$$

$$R_t^f = diag[R_t^1, R_t^2, \dots \dots R_t^m] \tag{21}$$

where g_t^f in (16) is the fused measurement vector, by combining the measurements of m sensors (ISNs) at time instant t. Similarly H_t^f is the corresponding value of the Jacobian of the respective ISNs. In (21) R_t^f is the measurement error covariance. Note that no modification measurements of the ISNs is carried out here and pure measurements of the target are being fused at the CHN to obtain the final state vector and state error covariance.

3.2.2 WF method

This method is based on combining the m measurements in a different manner than MKF. A weighing factor ϖ is allotted to each of the corresponding measurements of the ISNs, which represents the degree of correctness or confidence that one has regarding the measurement obtained from a specific ISN. The weight factor has been applied to Eqs. (19)–(21) as follows

$$g_t = \frac{\sum_{m=1}^{N} (\varpi_t^m g_t^m)}{\sum_{m=1}^{N} w_t^m} \tag{22}$$

$$H_t = \frac{\sum_{m=1}^{N} (\varpi_t^m H_t^m)}{\sum_{m=1}^{N} w_t^m} \tag{23}$$

$$R_t = \frac{\sum_{m=1}^{N} (\varpi_t^m R_t^m)}{\sum_{m=1}^{N} \varpi_t^m} \tag{24}$$

where g_t, H_t and R_t are the composite measurement vector, measurement-matrix/Jacobian and measurement noise covariance matrix respectively obtained by combining respective components from the m sensors sensing the target at that specific time instant. The ϖ_t^m is the weight alotted to the mth sensor at t^{th} time instant. Possible choices for the weights are

$$\varpi_t^m = \frac{1}{R_t^m} \tag{25}$$

$$\varpi_t^m = \frac{1}{(d_t^m)^r} \tag{26}$$

where R_t^m represents the measurement noise of the m^{th}sensor, d_t^m the distance of the mth sensor from the target and r represents the path loss exponent. In our simulations we have used Eq. (25). We utilize the above weighted - fused quantities in the EKF and the CGKF.

3.3 Heterogeneous DF

Heterogeneous DF differs from the homogeneous variety in that we fuse data from different types of sensors in combinations: such as, PIR and acoustic or PIR and seismic or PIR, acoustic and seismic together [12]. We have tried the architectures of centralized (measurement fusion) as well as decentralized (state fusion) data fusion. There are several methods in practice for DF but for nonlinear measurement models, it has been found that only a few models have been able to maintain the accuracy against catastrophic fusion [23]. The state model applied is the DWNA (Eq. (6)) of the subsection A.

3.3.1 Centralized DF

This architecture mainly follows the measurement fusion. The measurements (data) are obtained from all ISNs and then fused at cluster head node CHNs. In our case the data obtained is nonlinear from all three sensors with different size of measurement models. The only possible approach to collate data effectively is the MKF since weighted fusion applies only to sensors based on similar measurement model. The method has been applied to both EKF and CGKF.

The MKF is an effective way to combine data from dissimilar type of sensor measurement models. At the cost of computational complexity owing to matrix size this is overall an effective method considering WF can combine data from only similar group of sensors.

3.3.2 Decentralized data fusion

The method has been explicitly used to bring out the fact the CGKF did perform better as against any of these methods of combining state parameters and covariances. This method has been cited as state fusion concept [12] or hierarchical data fusion. This is based on a two tier system wherein state estimation of the target is carried out at ISNs which forms tier 1 and these states are then fused at tier 2 in the CHNs. The global state estimate and global state error covariance calculated at CHNs and these are then fed to ISNs. The KF algorithm runs in the ISN to obtain fresh state and error covariance estimates, which are again fed at the CHN and the cycle continues. There are mainly two approaches of track to track fusion as given by Raol [12] in Eq. (27) and (28) and Durrant whyte [24] in Eqs. (29) and (30). Most of the methods surveyed in this category feature a scheme where in we have to combine error covariances and state vectors to produce new covariance and state vectors. The only difference between the two methods below is the way state estimates and error covariances are used to compute fused global values of state estimate $\overline{X}^f_{t+1}$ and state error covariance $\overline{P}^f_{t+1}$ at $t+1$ time instant. The symbols used in the equations below $P^1$and P^2 are the covariances with respect to two different sensors. $\hat{X}^1_t$ and $\hat{X}^2_t$ are the target state vector as computed by the EKF at ISNs. $\overline{X}^f_{t+1}$ and $\overline{P}^f_{t+1}$ are the finally computed target state vector and state error covariances respectively at CHNs. This is fed back to every ISN after every iteration. In the present work we use the Global fusion method of Raol [12].

Global Fusion

$$\overline{X}^f_{t+1} = \hat{X}^1_t + \hat{P}^1_t\left(\hat{P}^1_t + \hat{P}^2_t\right)^{-1}\left(\hat{X}^2_t - \hat{X}^1_t\right) \tag{27}$$

$$\overline{P}^f_{t+1} = P^1_t - P^1_t\left(\hat{P}^1_t + \hat{P}^2_t\right)^{-1}P^{1T}_t \tag{28}$$

Track to Track Fusion

$$\overline{P}^f_{t+1} = \left[\sum_{i=1}^{N} P^{-1}_{i_t}\right]^{-1} \tag{29}$$

$$\overline{X}^f_{t+1} = \overline{P}^f_{t+1}\sum_{i=1}^{N} P^{-1}_{i_t}X_{i_t} \tag{30}$$

3.4 Maneuvering target

The class of maneuvering targets yield particularly challenging tracking problems. The challenges include choosing a system model close to the actual target maneuvers in addition to often having to give real time solutions. In our work we now aim to demonstarate the efficacy of the CGKF framework to RADAR- measurement based coordinated turn (CT) models. We reiterate that the non necessity of prior knowledge of the system and measurement noise characteristics (often representing the nature of maneuver) make the CGKF particularly attractive. The present work builds on [15] where the CGKF algorithm has been applied to a variety of maneuvering targets based on a linear measurement model. Currently a non linear measurement model (RADAR based) has been employed in order to move a step closer to a more realistic scenario. We have applied the CGKF to the highly maneuvering class of CT models with known as well as unknown turn rates [13, 14]. In the simulation studies the turn rate is represented by ω.

The present part is divided into the following parts.

3.4.1 CT state variable model

A two dimensional model for the target tracking problem (maneuver in horizontal 2D plane) is described as follows.

State Equation: CT known ω.

$$X_{t+1} = AX_t + Bw_t \tag{31}$$

where state vector is $X_t = (x(t)\ \dot{x}(t)\ y(t)\ \dot{y}(t))^T$, state transition matrix $A = \begin{pmatrix} A_1 & -A_2 \\ A_2 & A_1 \end{pmatrix}$, $B = \begin{pmatrix} B_1 & B_2 \\ B_2 & B_1 \end{pmatrix}$.

where $A_1 = \begin{pmatrix} 1 & Sin(\omega\Delta t)/\omega \\ 0 & Cos(\omega\Delta t) \end{pmatrix}$, $A_2 = \begin{pmatrix} 0 & (1 - Cos(\omega\Delta t))/\omega \\ 0 & Sin(\omega\Delta t) \end{pmatrix}$,

$B_1 = \begin{pmatrix} \Delta t^2/2 & 0 \\ \Delta t & 0 \end{pmatrix}$, $B_2 = \begin{pmatrix} 0 \\ 0 \end{pmatrix}$ and w_t represents system noise which is Gaussian and Δt is a time step. Here we have considered the state vector to include X and Y coordinates of the target as well as the speed in the two coordinates.

State Equation: CT with unknown ω

$$X_{t+1} = A(X_t)X_t + Bw_t \tag{32}$$

A two dimensional model for the target tracking problem (maneuver in horizontal 2D plane) is described as follows.

where state vector is $X_t = (x(t) \quad \dot{x}(t) \quad y(t) \quad \dot{y}(t) \quad \omega(t))^T$, state transition matrix $A(X_t) = \begin{pmatrix} A_1 & -A_2 & B_2 \\ A_2 & A_1 & B_2 \\ B_3 & B_3 & 1 \end{pmatrix}$, $B = \begin{pmatrix} \Delta t^2/2 & 0 & 0 \\ \Delta t & 0 & 0 \\ 0 & \Delta t^2/2 & 0 \\ 0 & 0 & \Delta t \\ 0 & 0 & \Delta t \end{pmatrix}$ where $B_3 = (0 \quad 0)$, A_1,A_2,B_2,B_3 are as described above and w_t represents system noise which is white Gaussian. Inclusion of the angular speed ω in the state vector makes the state equation non linear for thecase of CT with unknown ω.

Measurement Equation:

$$g_t = \begin{pmatrix} \left(x(t)^2 + y(t)^2\right)^{.5} \\ tan^{-1}(y(t)/x(t)) \end{pmatrix} + v_t \tag{33}$$

where g_t is the measurement vector and v_t is measurement noise which is assumed to be white Gaussian.

4. Results and sensitivity studies

4.1 Stand alone mode

The tabulated result of all sensors for EKF and CGKF are given below with their respective PFE (Percentage Fit Error). The error metric $PFE = \frac{|X_t - \overline{X}_t|}{|X_t|} \times 100$ which represents the normalized difference between the estimated and actual track, achieved by CGKF and EKF. The PIR sensor gives the least error with CGKF. All the results in this section and subsequent sections are out of a minimum of 500 Montecarlo runs. The plots for EKF (Left) and CGKF (Right) have been combined together. The figures appear as top and bottom, top one is the true trajectory and bottom one is the true trajectory super imposed with estimated trajectory. The PFE is also mentioned on the graph itself for every case. This is same for all the cases given below.Where ever the error is negligible, the estimated track (green) completely takes over the actual track (black). Following are the deductions based on the simulation results. It is to be noted that the plots are based on one of the 500 runs used to compute the PFE metric (refer **Table 2**). This applies to the present and all subsequent sections also.One example of each sensor performance is displayed in the **Figures 2–4** respectively. The PFE, RMSPE metric (representative of the error in range calculation based on x, y coordinates of the target and expresssed as) in the plots correspond to that of the CGKF for a particular run.

Sensor Type	No of Sensors	EKF (PFE) %	CGKF (PFE) %
PIR	1	3.77076	1.03723
Acoustic	1	2.497723	1.9393
Seismic	1	4.622587	2.614668

Table 2.
Stand alone mode.

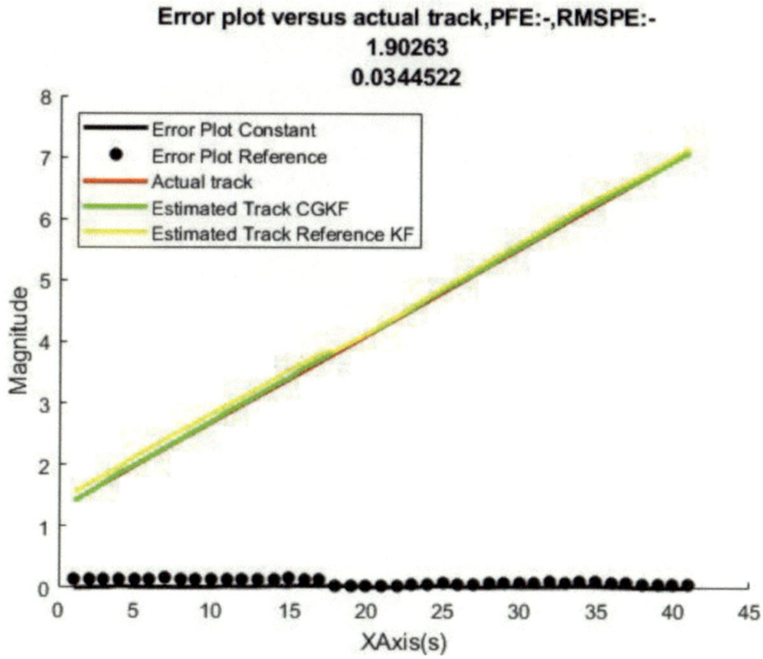

Figure 2.
Stand alone mode:-PIR sensor.

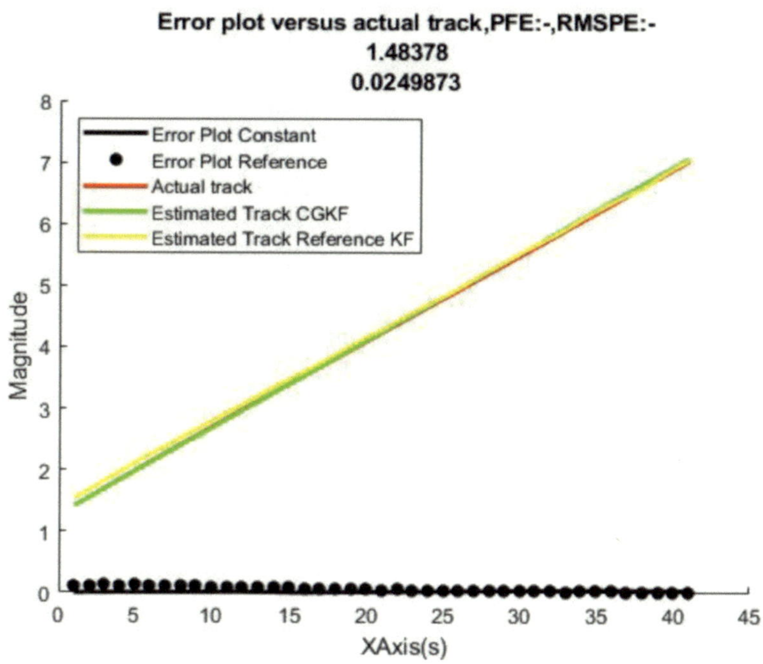

Figure 3.
Stand alone mode:-acoustic sensor.

4.2 Homogeneous fusion mode of sensors

The results from both, MKF and Weighted fusion have been tabulated seperately as shown in the first six enteries of **Table 3**. Settings of the simulations

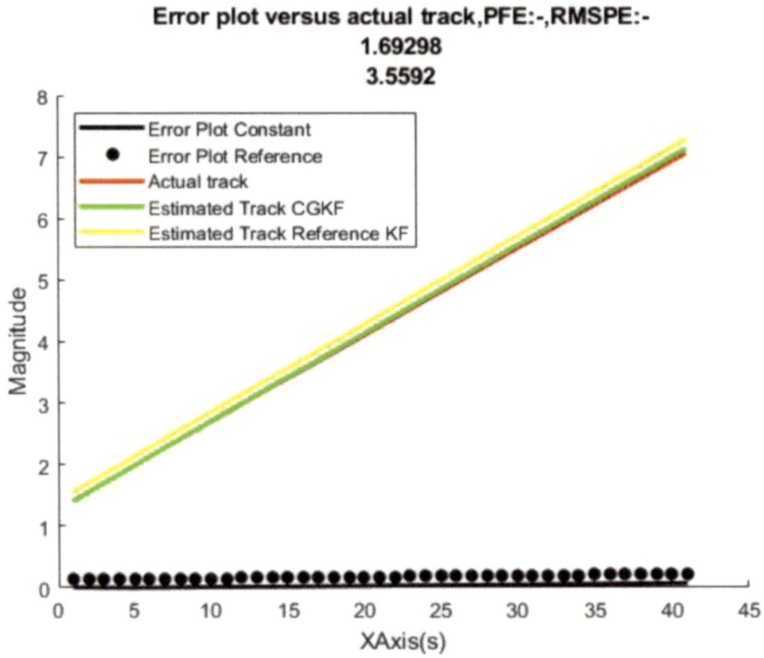

Figure 4.
Stand alone mode:- Sesimic sensor.

Sensor Type	EKF (PFE) %	CGKF (PFE)%	Fusion Type
PIR	6.32816	2.81109	Maximal
Acoustic	5.76208	5.61651	Maximal
Seismic	5.54418	2.38094	Maximal
PIR	0.580869	0.579842	Homogeneous Weighted
Acoustic	1.47602	1.33768	Homogeneous Weighted
Seismic	1.7170850	1.0956019	Homogeneous Weighted
PIR & Acoustic	18.5708	9.3724	Maximal
PIR, Acoustic & Seismic	19.9958	1.34023	Maximal
PIR & Seismic	10.9576	3.42436	Maximal

Table 3.
Measurement fusion:-4 sensor set.

including the number of Monte Carlo runs is same as that for the Stand Alone method described above. An example of the method is illustrated in **Figures 5** and **6**. The PFE, RMSPE metrics (as defined in Section 4.1) metric in the plots correspond to that of the CGKF for the coresponding run.

4.3 Heterogeneous fusion mode of sensors

The results in last three entries of **Table 3**, are those corresponding to the measurement fusion based method of heterogeneous fusion. Settings of the simulations including the number of Monte Carlo runs is same as that for the Stand Alone and homogeneous fusion method described above. One example each

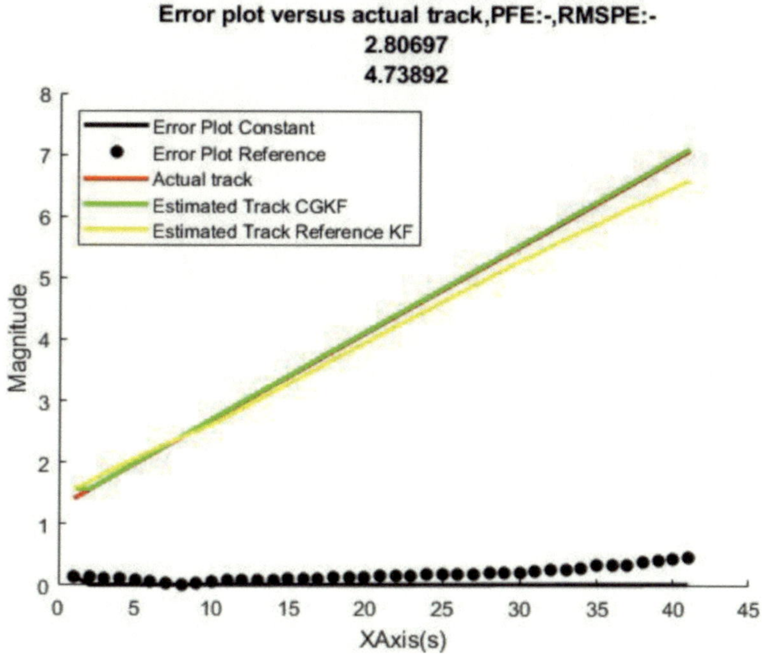

Figure 5.
Homogeneous fusion (MKF):- seismic sensor.

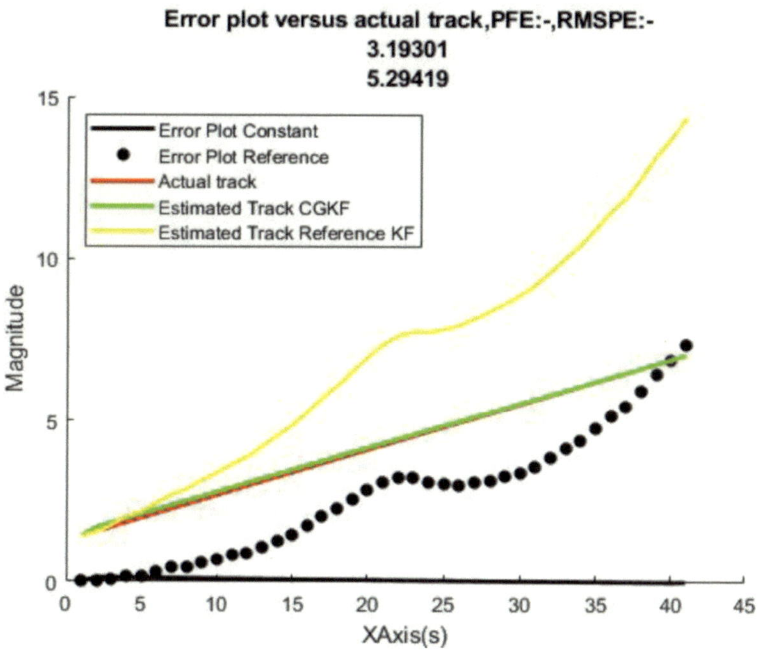

Figure 6.
Homogeneous fusion (weighted):- acoustic sensor.

of the two sensor (PIR and seismic case) and three sensor (all three combined) is illustrated in **Figures 7** and **8** repectively. The PFE, RMSPE metrics (as defined in Section 4.1) in the plots correspond to that of the CGKF for the coresponding run (refer **Table 4**).

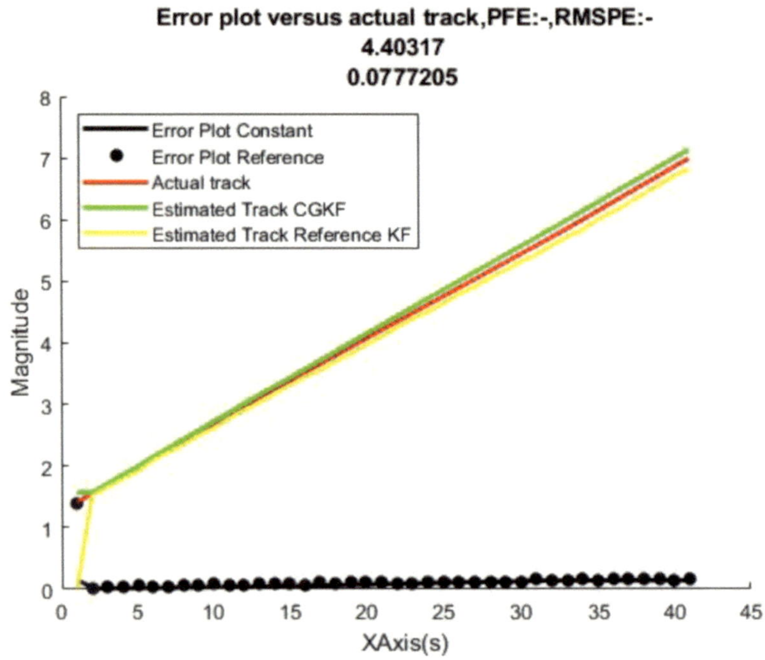

Figure 7.
Heterogeneous fusion (maximal):-PIR and seismic sensors.

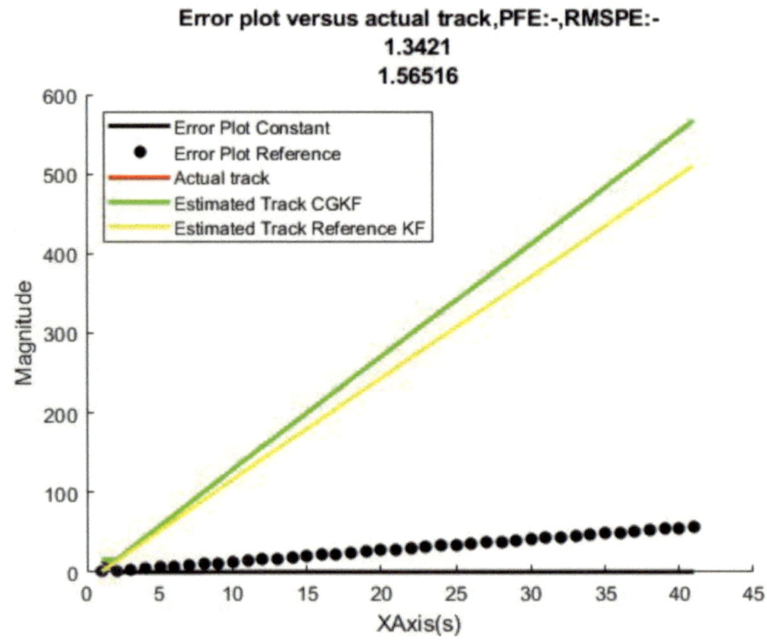

Figure 8.
Heterogeneous fusion (maximal):- PIR, seismic and acoustic sensors.

4.4 Maneuvering target

The 2-D frame work study has been carried out on a set of seventy data points in order to generate a smooth trajectory. The following system and measurement covariances matrices are used to generate the simulated track $Q = .01I$, $R = .1I$ for

Sensor Type	Fusion Type(State-Fusion)	EKF (PFE)%
PIR & Acoustic	Global Fusion [12]	2.29092
PIR, Acoustic & Seismic	Global Fusion [12]	NaN(Not a Number)*
PIR, Acoustic & Seismic	Whyte Method [24]	1.1158

*:- *Due to lack of convergence.*

Table 4.
Heterogeneous fusion (state fusion).

all models, choice of the initial value of ω in the CT model has been obtained by via standard fighter aircraft (eg F-16) data available on the internet. This value of ω has been set to .5, which corresponds to approx 28.6 degrees/s (which happens to be the maximum instantaneous turn rate of any current generation fighter aircraft). Here the *PFE* is based on sum total *PFE* obtained along X and Y coordinates respectively. The error metric shown in tables is the average value computed over 500 runs while the plots correspond to one specific run wherein results are presented in the form of 2D plots of the simulated target trajectory, simulated measurements and the estimated track against time. **Table 5** shows the typical constant gain average values computed using GA over 500 runs corresponding to each of the models. Corresponding to a certain gain there is a transient and steady state behavior. If the gain is large the transient is short with the steady state fluctuating error being large. When the gain is small as in the present case there is a large transient with small steady state error. If the filter is run backwards from the end then the whole actual trajectory will be wrapped around by the estimated values. In a nutshell the filter gain values K, can be tuned manually to provide optimal tracking results in a constant gain framework. The filter gain values will be of typical nature as per **Table 5** corresponding to specific target state models. **Figures 9** and **10** illustrate the performance of the CGKF versus the standard KF model. The PFE, RMSPE metrics (as defined in Section 4.1) in the plots correspond to that of the CGKF for a particular run.

4.5 Sensitivity studies on constant gain in case of maneuvering targets (CT (known ω))

Under this heading we demonstrate the robustness of the constant gain in so far as the application of gain variations to the maneuvering target tracking scenario for

MODEL	EKF %	CGKF %	K matrix
CT(known ω)	56.75	10.11	$\begin{pmatrix} .0005 & .0013 \\ .0013 & .0013 \\ .0013 & .0013 \\ .0013 & .0013 \end{pmatrix}$
CT(unknown ω)	24	14.78	$\begin{pmatrix} .0013 & .00007 \\ .0012 & 0 \\ 0 & .0013 \\ .00007 & .0013 \\ .0013 & .0004 \end{pmatrix}$

Table 5.
Percentage fit error comparison:-non linear case and typical K matrix values.

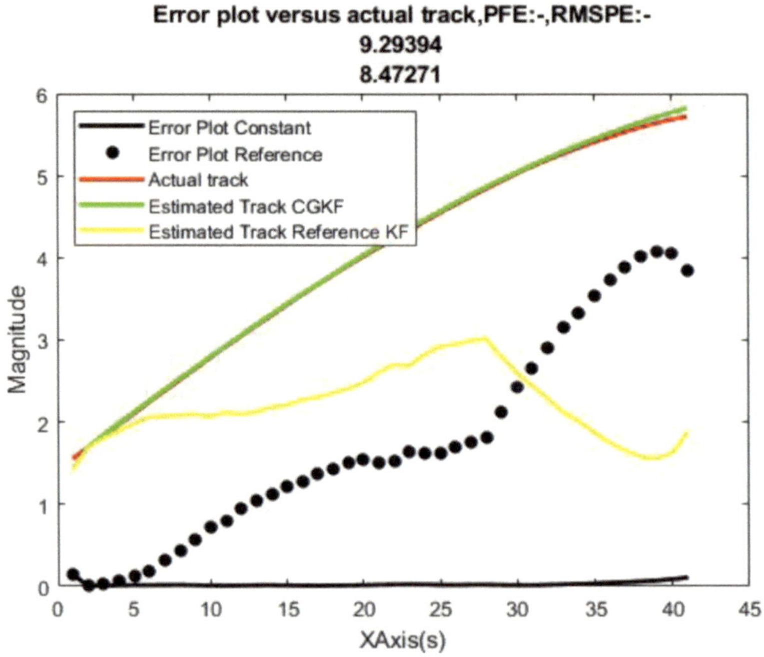

Figure 9.
CT(known ω).

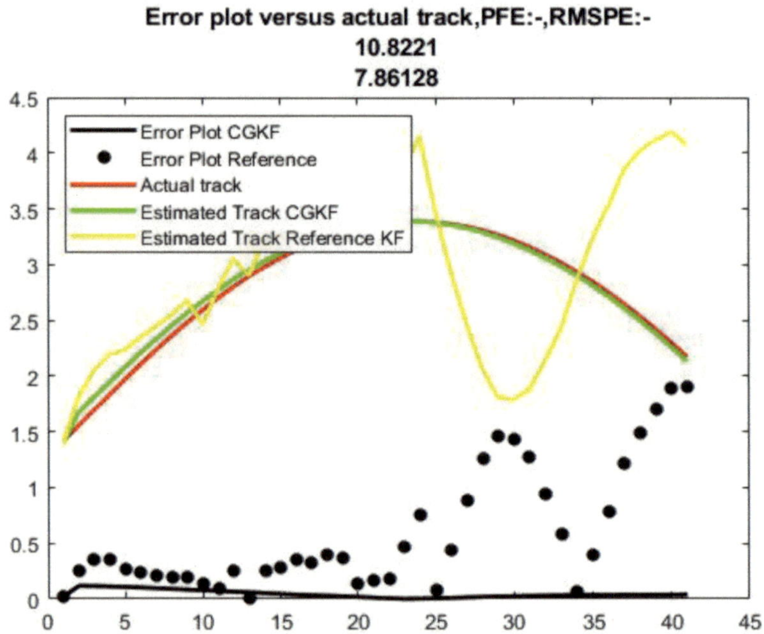

Figure 10.
CT(unknown ω).

the CT (known ω) is concerned. In the tables below are mentioned different PFE/RMSPE metrics achieved as per specified variation in the constant gain values are concerned. In **Table 6** we show variation as per additive increments to the constant gain while in **Table 7** we show variation as per fractional values to the constant gain.

Additive variations (K)	PFE%	RMSPE%
K(1 + .1 randn)	10.9506	13.3848
K(1 + .2 randn)	9.02807	9.8478
K(1 + .3 randn)	11.4343	13.3211
K(1 + .4 randn)	11.4565	13.77
K(1 + .5randn)	8.68005	9.92213
K(1 + .6 randn)	10.3575	10.5219
K(1 + .7 randn)	13.1419	18.0384
K(1 + .8 randn)	10.4037	12.4931

Table 6.
Constant gain robustness to additive variations (CT(known ω)).

Fractional Variations (K)	K/8	K/4	K/2	2 K	4 K	8 K
PFE%	11.1569	9.34031	12.7851	13.0054	11.6087	9.11502
RMSPE%	14.2031	11.0111	15.2803	15.1227	13.0654	10.1699

Table 7.
Constant gain robustness to fractional variations (CT(known ω)).

The tables show that the constant gain is robust to minor additive and fractional increments, thereby demonstrative of the fact that the achieved constant gain provides good tracking results as far as achieved PFE values indicate.

5. Conclusion

We believe that these are the only studies of a CGKF applied to tracking targets in WSN environments and maneuvering target models based on non linear measurement models. As seen the EKF is unable to effectively track the targets in WSN and for the maneuvering target case compared to the CGKF. This is a significant finding and supports the fact that CGKF effectively circumvents, or in other words trades the gains with the filter statistics which are more difficult to obtain and therein gives optimal tracking results by working directly with the Kalman Gain. The present results prove that the CGKF is successful in target tracking applications wherein the constant gain approach overcomes uncertainty regarding noise statistics that exist in the framework of the problem. The CGKF has been employed for tracking maneuvering targets and those in a WSN. The present work firmly establishes the CGKF framework thereby enabling its applicability to a wider variety of problems as deemed fit by the reader.

5.1 Analysis of results and future work

5.1.1 Stand alone mode

Following are the deductions based on the simulation studies as summarized in **Table 2**.

1. The results and plots bring out clearly the novelty of CGKF, the overall performance of which is better than the EKF as per the PFE values.

2. In the case of the EKF the Acoustic sensor performs the best.

3. In case of the CGKF the PIR performs the best.

5.1.2 Homogeneous fusion mode

Following are deductions based on the simulation studies as summarized in **Table 3** [16].

1. The overall performance of the CGKF is better than the EKF for both the MKF and Weighted methods.

2. Considering the CGKF case the PIR and seismic sensors peroform better than the acoustic sensor, with the PIR performing the best overall.

3. Amongst the various fusion methods the overall performance of the weighted fusion is better compared to the MKF, for all types of sensors.

5.1.3 Heterogeneous Fusion mode

Following are the deductions based on the simulation studies as summarized in **Table 4**.

1. Overall the CGKF performance is better than the EKF for heterogeneous fusion method.

2. With reference to heterogeneous fusion of PIR, acoustic and seismic sensors the Durrand Whyte method [24] gives better results compared to Global fusion method [12]. Here we note that a comparison with the CGKF is not possible since the CGKF works with purely measurements and not by propagation of state error covariances which is fundamental to these techniques. The Global fusion method [16] does not provide convergence in tracking when using PIR, acoustic and seismic sensors together.

3. The CGKF heterogeneous fusion model of PIR, Acoustic and seismic sensors gives optimum performance better than its EKF counterpart.

4. PIR based weighted fusion gives better results than the heterogeneous fusion. However we must keep in mind the fact that the simulations for heterogeneous fusion are based only one sensor of each type unlike the homogeneous fusion case where four sensor of each type are considered. The plots and result have been mentioned under. All the result has been obtained through Montecarlo simulation with runs of average of 500.

5.1.4 Maneuvering target

Following are the deductions of the simulations.

1. **Figures 9** and **10** and **Table 5**, clearly show that the performance of the CGKF is very much better than that of the EKF.

2. The results obtained show the CGKF performing better than the EKF in three models (ie. DWPA and both CT models).

DOI: http://dx.doi.org/10.5772/intechopen.98700

3. The results obtained for the CT models including those of sensitivity analysis (Refer **Tables 6** and 7) demonstrates the viability of applying the CGKF to this category of problems.

5.2 Conclusions and Suggestions for Further Studies

The efficacy of the CGKF has been demonstrated wherein a single approach yields optimal results for a varierty of linear [10] as well as non linear models in WSN and maneuvering target scenarios [15]. The extensive numerical studies establish the fact that the CGKF performs better that the conventional EKF.

Actual implementation of a target tracking application in the WSN environment shall require optimal routing, deployment, design, communication protocols and other such associated integral characteristics mentioned in the introduction. Though not directly within the purview of the scope of the work, these aspects are very important.

It would be very useful to apply this CGKF to variants of the Kalman Filter such as particle filter, ensemble filter and other formulations.

Finally CGKF could be tried out for massive data based problems like numerical weather prediction. The constant gains can be pre computed using earlier data and since the gains are robust they can be expected to handle newer data quiet efficiently similar to space debris as in [1].

Author details

Peeyush Awasthi[1], Ashwin Yadav[2]*, Naren Naik[3] and Mudambi Ramaswamy Ananthasayanam[4]

1 Graduate Research Scholar, Florida International University, USA

2 Geomatics Engineering, Department of Civil Engineering, Indian Institute of Technology, Roorkee, India

3 Department of Electrical Engineering, Indian Institute of Technology Kanpur, India

4 Department of Aerospace Engineering, Indian Institute of Science, Bangalore, India

*Address all correspondence to: ashwiny77@gmail.com

References

[1] Mudambi R. Ananthasayanam (November 23rd 2018). Tuning of the Kalman Filter Using Constant Gains, Introduction and Implementations of the Kalman Filter, Felix Govaers, IntechOpen, DOI: 10.5772/ intechopen.81795. Available from: https://www.intechopen.com/books/ introduction-and-implementations-of-the-kalman-filter/tuning-of-the-kalman-filter-using-constant-gains

[2] Mehra, R.K. , “Approaches to adaptive filtering,” in *IEEE Symposium on Adaptive Processes (9th) Decision and Control* , vol.9, no., pp.141, 7-9 Dec. 1970

[3] Robert H Shumway, David S Stoffer; “Time series Analysis and its applications” (Springer Text in Statistics)

[4] Bavdekar, V. A., Deshpande, A. P. and Patwardhan, S. C. (2011) Identification of process and measurement noise covariance for state and parameter estimation using extended Kalman filter. Journal of Process control, 21 : 585-601.

[5] A. H. Mohamed, K. P. Schwartz, “Adaptive Kalman filtering for INS/ GPS,” *Journal of Geodesy*, vol 73(2), pp. 193-203,1999.

[6] Myers, K.; Tapley, B.; , “Adaptive sequential estimation with unknown noise statistics,” , *IEEE Transactions on Automatic Control* , vol.21, no.4, pp. 520-523, Aug 1976

[7] R.M.O Gemson and M.R. Ananthasayanam, “Importance of Initial State Covariance Matrix for the parameter estimation using an Adaptive Extended Kalman Filter”,in American Institute of Aeronautics and Astronautics, vol. 4153, pp. 94-104,1998.

[8] Akita, T.; Takaki, R,; Shima, E, “A new adaptive estimation method of spacecraft thermal mathematical model with an ensemble Kalman filter”, Acta Astronautica, vol. 73, April-May 2012, pp 144-155

[9] A.K. Anil Kumar,M.R. Ananthasayanam,P.V. Subba Rao, “A Constant Gain Kalman Filter Approach for the prediction of the re-entry of risk objects”,in *Acta Astronautica* , vol 61 (10), vol-25,pp. 831-839,2007.

[10] Yadav, A.; Naik, N.; Ananthasayanam, M. R.; Gaur, A.; Singh, Y. N., " A constant gain Kalman filter approach to target tracking in wireless sensor networks," Industrial and Information Systems (ICIIS), 2012 7th IEEE International Conference on , vol., no., pp.1,7, 6-9 Aug. 2012

[11] Christian Bohn., “Recursive Parameter Estimation Of Non Linear Continuous-Time Systems through Sensitivity-Model-Based Adaptive Filter”,2000

[12] Jitendra.R.Raol, *Multi Sensor Data Fusion with MATLAB*, CRC Press,2010.

[13] X. Rong Li; Vesselin P Jilkov; , “A Survey of Maneuvering Target Tracking: Dynamic Models”, SPIE conference on Signal and Data processing of small targets, April 2000 (4048-22)

[14] Yaakov Bar-Shalom; Peter K Willet; Xin Tian,*Tracking and datafusion A handbook of algorithms,* YBS Press,2011.

[15] Yadav, A.; Awasthi, P.; Naik, N.; Ananthasayanam, M.R., " A constant gain Kalman filter approach to track maneuvering targets," Control Applications (CCA), 2013 IEEE International Conference on , vol., no., pp.562,567, 28-30 Aug. 2013

[16] R.E Kalman, "A new approach to Linear Filtering and Prediction Problems", in *Transactions of the ASME-Journal of Basic Engineering*, vol. 82, pg-35-45., 1960

[17] Mehra, R.K., " On the Identification of variances and and adaptive Kalman filtering," Automatic Control, IEEE Transactions on , vol. AC-15, No. 2, April 1970

[18] Kailath, T., " An innovations approach to least-squares estimation– Part I: Linear filtering in additive white noise," Automatic Control, IEEE Transactions on , vol.13, no.6, pp.646,655, Dec 1968

[19] Viswanath. Study of constant gain Kalman filter for basic systems with preliminary study of ISAR problem [MTech thesis]. India: IIT Kanpur; 2018

[20] Kalyanmoy Deb, *Multi-Objective Optimization using Evolutionary Algorithms*, Wiley India , 2010.

[21] Nithya V.S.;Sheshadri K; Kumar A; Hari K.V.S; "Model based target tracking in a wireless network of passive infrared sensor nodes" International Conference on Signal Processing and Communications (SPCOM), 2010, 18-21 July 2010

[22] Stefano Coraluppi, Craig Carthel, Mahendra Mallick; "Multi-Target tracking with Unattended Ground Sensors (UGS) Data; ALPHATECH,Inc

[23] H.B.Mitchell; Multi-Sensor Data Fusion, An Introduction 2007

[24] Hugh Durrant-Whyte Lecture notes on "Multi Sensor Data Fusion"

[25] Grimble M J, Jukes K A and D P Goodall, " Nonlinear filters and operators and the constant gain extended Kalman filter," *IMA Journal of Mathematical Control and Inf.*, vol. 1, pp. 359-386, 1984.

Section 3

Practical Implementations and Applications

Chapter 5

Parameter Estimation of Weighted Maxwell-Boltzmann Distribution Using Simulated and Real Life Data Sets

Javaid Ahmad Reshi, Bilal Ahmad Para and Shahzad Ahmad Bhat

Abstract

This paper deals with estimation of parameters of Weighted Maxwell-Boltzmann Distribution by using Classical and Bayesian Paradigm. Under Classical Approach, we have estimated the rate parameter using Maximum likelihood Estimator. In Bayesian Paradigm, we have primarily studied the Bayes' estimator of the parameter of the Weighted Maxwell-Boltzmann Distribution under the extended Jeffrey's prior, Gamma and exponential prior distributions assuming different loss functions. The extended Jeffrey's prior gives the opportunity of covering wide spectrum of priors to get Bayes' estimates of the parameter – particular cases of which are Jeffrey's prior and Hartigan's prior. A comparative study has been done between the MLE and the estimates of different loss functions (SELF and Al-Bayyati's, Stein and Precautionary new loss function). From the results, we observe that in most cases, Bayesian Estimator under New Loss function (Al-Bayyati's Loss function) has the smallest Mean Squared Error values for both prior's i.e., Jeffrey's and an extension of Jeffrey's prior information. Moreover, when the sample size increases, the MSE decreases quite significantly. These estimators are then compared in terms of mean square error (MSE) which is computed by using the programming language R. Also, two types of real life data sets are considered for making the model comparison between special cases of Weighted Maxwell-Boltzmann Distribution in terms of fitting.

Keywords: Weighted Maxwell-Boltzmann Distribution, prior distributions, loss functions, R Software

1. Introduction

In Statistical Mechanics, there are a lot of applications of Maxwell-Boltzmann Distribution. The Maxwell-Boltzmann distribution forms the basis of the kinetic energy of gases, which explains many fundamental properties of gases, including pressure and diffusion. This distribution is sometimes called as the distribution of velocities, energy and magnitude of momenta of molecules. Tyagi and Bhattacharya [1] who considered the Maxwell distribution as a lifetime model and discussed the Baye's and minimum variance unbiased estimation procedures for its parameter and reliability function. Chaturvedi and Rani [2] estimated the classical and Baye's estimators for the Maxwell distribution, after generalization it by adding another parameter. Empirical Baye's estimation for the Maxwell distribution was also obtained by Bekker and Roux [3]. Kazmi et al. [4] derived the Bayesian estimation for two component mixture of Maxwell distribution, assuming censored data. The Maxwell-Boltzmann distribution can be used to find the distribution of particle's kinetic energy which is related to particle's speed by the formula $E = mv^2/2$, provided the distribution of speed is known. The PDF of Maxwell-Boltzmann distribution is given by Maxwell [5]:

$$f_w(x) = \sqrt{(2/\pi)}\,\theta^{3/2}x^2e^{-\theta\,x^2/2} \tag{1}$$

And the CDF of Maxwell Distribution is given as:

$$F_w(x) = 1 - \frac{\Gamma((\alpha+3)/2, \theta x^2/2)}{\Gamma((\alpha+3)/2)} \tag{2}$$

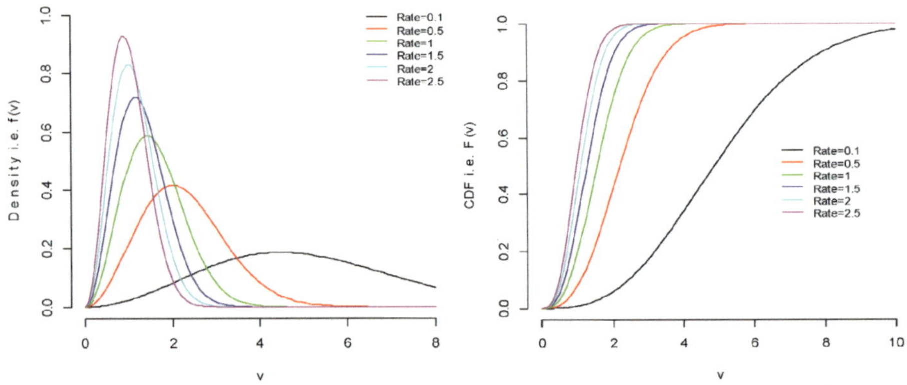

Recently, Aijaz et al. [6] estimates and analyze the Bayes' Estimators of Maxwell-Boltzmann Distribution under various Loss functions and prior Distributions. Other Contributions in Maxwell Distribution are Huang and Chen [7], Krishna and Malik [8], Tomer and Panwar [9], Zhang et al. [10], and Monisa [11].

Various Statisticians and Mathematicians have carried out the Bayesian paradigm of Maxwell-Boltzmann distribution by using loss functions and prior distributions, see Al-Baldawi [12], Dey et al. [13], Podder and Roy [14], Rasheed [15], and Spiring and Yeung [16].

The concept of weighted distributions introduced by Fisher [17] and later it was formulated in general terms by Rao [18] in connection with modeling statistical data. These Distributions are applicable, when each and every observation is given an equal chance of being recorded. These distributions arise, when the probability of selecting an observation varied from observation to observation. In this context, the authors generalize the Maxwell Distribution and is known as Weighted Maxwell-Boltzmann distribution. The PDF of Weighted Maxwell-Boltzmann Distribution was introduced by Aijaz et al. [19].

$$f_w(x;\theta,\alpha) = \frac{\theta^{(\alpha+3)/2}x^{(\alpha+2)}e^{-\theta\,x^2/2}}{2^{(\alpha+1)/2}\Gamma((\alpha+3)/2)} \tag{3}$$

Where θ is the rate parameter and ω is the weight parameter (ω>0).
Also, CDF of the Weighted Maxwell Distribution is given by:

$$F_w(x) = 1 - \frac{\Gamma((\alpha+3)/2, \theta x^2/2)}{\Gamma((\alpha+3)/2)} \tag{4}$$

The Reliability function and Hazard Rate of the Weighted Maxwell Distribution is given by:

$$R_w(x) = \frac{\Gamma\left((\alpha+3)/2, \theta x^2/2\right)}{\Gamma((\alpha+3)/2)} \tag{5}$$

$$h_w(x) = \frac{\theta^{((\alpha+3)/2)} x^{(\alpha+2)} \exp\left(-\theta x^2/2\right)}{2^{((\alpha+1)/2)} \Gamma((\alpha+3)/2) \Gamma\left((\alpha+3)/2, \theta x^2/2\right)} \tag{6}$$

The, r^{th} moments about zero of Weighted Maxwell-Boltzmann Distribution is given by:

$$\mu_r' = (2/\theta)^{\frac{r}{2}} \Gamma((\alpha+r+3)/2)/\Gamma((\alpha+3)/2), \text{ Where } r = 1, 2, 3, 4, \ldots \tag{7}$$

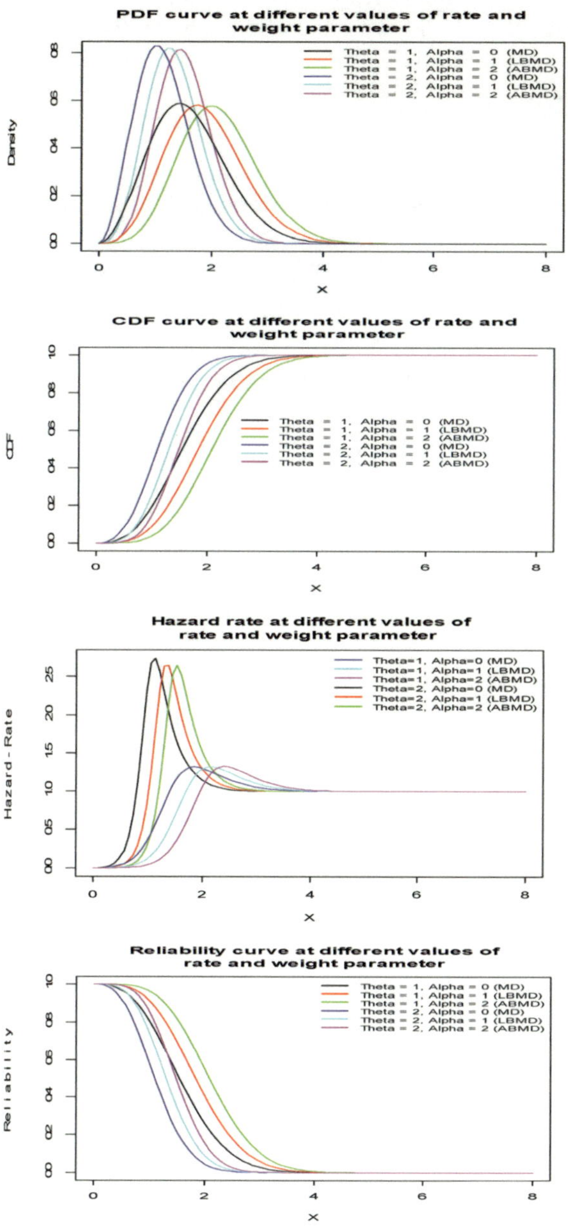

In comparison to classical approach, Bayesian approach is considered to be fair enough in estimating the parameters of a distribution provided that the prior distribution describes nicely the random behavior of a parameter. Very often, priors are chosen according to one's subjective knowledge and beliefs that is why Bayesian approach is sometimes called as subjective approach. However, Aslam [20] have shown an application of prior predictive distribution to elicit the prior density.
A number of symmetric and asymmetric loss functions have been shown to be functional, see Kasair et al. [21], Norstrom [22], Reshi et al. [23], Zellner [24], Reshi et al. [25], Dey and Maiti [26], Alkutbi [27], Wald [28], etc.

2. Estimation of parameters

In this Section, the authors estimated the parameters of Weighted Maxwell-Boltzmann Distribution under Classical and Bayesian Paradigm.

2.1 Maximum likelihood estimation

Let $x = (x_1, x_2, x_3, \dots, x_n)$ be a random sample of size n from Weighted Maxwell Distribution Therefore the likelihood function will be given by:

$$L(\theta, \omega/x) = \frac{\theta^{n\left(\frac{\omega+3}{2}\right)} \sum x_i^{\omega+2}}{2^{n\left(\frac{\omega+1}{2}\right)} \Gamma\left(\frac{\omega+3}{2}\right)^n} \exp\left(-\frac{\theta}{2}\sum_{i=1}^{n} x_{i^2}\right) \tag{8}$$

The, the Log likelihood function is given by:

$$\log L(\theta, \omega/x) = \frac{n(\omega+3)}{2}\log(\theta) - \frac{n(\omega+1)}{2}\log(2) - n\log\Gamma\left(\frac{\omega+3}{2}\right) + (\omega+2)\sum_{i=1}^{n}\log x_i - \frac{\theta}{2}\sum_{i=1}^{n} x_{i^2} \tag{9}$$

After, differentiating the log likelihood w.r.to θ, and equate to zero, we have:

$$\hat{\theta}_{\text{mle}} = \frac{n\omega + 3n}{\sum_{i=1}^{n} x_{i^2}} \tag{10}$$

2.2 Bayesian Estimation of Weighted Boltzmann Maxwell Distribution using different loss functions

2.2.1 Estimation using extension of Jeffery's prior

The Joint Probability Density Function of θ and x is given by:

$$f_1(x, \theta) = \frac{\theta^{\frac{n\omega+3n}{2} - 2c_1} \sum x_i^{\omega+2}}{2^{n\left(\frac{\omega+1}{2}\right)} \Gamma\left(\frac{\omega+3}{2}\right)^n} \exp\left(-\frac{\theta}{2}\sum x_{i^2}\right) \tag{11}$$

And, the marginal distribution function of θ and x is given by:

$$f_1(x) = \frac{\sum x_i^{\omega+2}}{\Gamma\left(\frac{\omega+3}{2}\right)^n} \frac{\Gamma\left(\frac{n\omega+3n-4c_1+2}{2}\right) 2^{\frac{2n-4c_1+2}{2}}}{\left(\sum x_{i^2}\right)^{\frac{n\omega+3n-4c_1+2}{2}}} \tag{12}$$

Substituting the above two Eqs. (11) and (12), we get the Posterior Probability Density function of θ and x is given by:

$$\pi_1(\theta/x) = \frac{\theta^{\frac{n\omega+3n-4c_1}{2}}\left(\frac{\sum x_{i^2}}{2}\right)^{\frac{n\omega+3n-4c_1+2}{2}}}{\Gamma\left(\frac{n\omega+3n-4c_1+2}{2}\right)} \exp\left(-\frac{\theta}{2}\sum x_{i^2}\right) \tag{13}$$

2.2.1.1 Bayes' estimator under squared error loss function

The Risk Function Under SELF is given as:

$$R_{(\mathrm{Sq,EJ})}\left(\hat{\theta}\right) = c\hat{\theta}^2 + \frac{4c\left(\frac{n\omega+3n-4c_1+4}{2}\right)\left(\frac{n\omega+3n-4c_1+2}{2}\right)}{\left(\sum x_{i^2}\right)^2.} - \frac{4c\hat{\theta}\left(\frac{n\omega+3n-4c_1+2}{2}\right)}{.\left(\sum x_{i^2}\right)} \tag{14}$$

After solving the above risk function, we get the Baye's estimator:

$$\hat{\theta}_{(\mathrm{Sq,EJ})} = \frac{n\omega+3n-4c_1+2}{\left(\sum x_{i^2}\right)} \tag{15}$$

2.2.1.2 Baye's estimator under precautionary: Loss function

The Risk Function Under Precautionary Loss Function is given as:

$$R_{(\mathrm{pr,EJ})}\left(\hat{\theta}\right) = c\hat{\theta} + \frac{4c\left(\frac{n\omega+3n-4c_1+4}{2}\right)\left(\frac{n\omega+3n-4c_1+2}{2}\right)}{\hat{\theta}\left(\sum x_{i^2}\right)^2.} - \frac{4c\left(\frac{n\omega+3n-4c_1+2}{2}\right)}{\left(\sum x_{i^2}\right)} \tag{16}$$

After solving the above risk function, we get the Baye's estimator:

$$\hat{\theta}_{(\mathrm{pr,Ej})} = \frac{\sqrt{(n\omega+3n-4c_1+4)(n\omega+3n-4c_1+2)}}{\left(\sum x_{i^2}\right)^2} \tag{17}$$

2.2.1.3 Baye's estimator under the Al-Bayyati's loss function

The Risk function under Al-Bayyati's Loss Function is given as:

$$R_{(\mathrm{Al,EJ})}\left(\hat{\theta}\right) = \frac{2^{c_2}\hat{\theta}^2\,\Gamma\left(\frac{n\omega+3n-4c_1+2c_2+2}{2}\right)}{\Gamma\left(\frac{n\omega+3n-4c_1+2}{2}\right)\left(\sum x_{i^2}\right)^{c_2}.} + \frac{2^{(c_2+2)}\Gamma\left(\frac{n\omega+3n-4c_1+2c_2+6}{2}\right)}{\Gamma\left(\frac{n\omega+3n-4c_1+2}{2}\right)\left(\sum x_{i^2}\right)^{(c_2+2)}} - \frac{2^{(c_2+2)}\hat{\theta}\,\Gamma\left(\frac{n\omega+3n-4c_1+2c_2+4}{2}\right)}{\Gamma\left(\frac{n\omega+3n-4c_1+2}{2}\right)\left(\sum x_{i^2}\right)^{(c_2+1)}} \tag{18}$$

After solving the above risk function, we get the Baye's estimator:

$$\hat{\theta}_{(\mathrm{Al,EJ})} = \frac{(n\omega+3n-4c_1+2c_2+2)}{\left(\sum x_{i^2}\right)} \tag{19}$$

2.2.1.4 Baye's estimator under the combination of Stein's loss function

The Risk function under the Stein's Loss function is given as:

$$R_{(\mathrm{St,Ej})}(\hat{\theta}) = \frac{\hat{\theta}\left(\frac{\sum x_{i^2}}{2}\right)}{\frac{n\omega+3n-4c_1}{2}} - (\log \hat{\theta}) + e^{-t} - 1 \tag{20}$$

After solving the above risk function, we get the Baye's estimator:

$$\hat{\theta}_{(\mathrm{St,Ej})} = \frac{n\omega + 3n - 4c_1}{\left(\sum x_{i^2}\right)} \tag{21}$$

2.2.2 Bayesian estimation under gamma (α, β) prior distributions

The Joint Probability Density Function of Maxwell-Boltzmann Distribution Using Gamma Prior Distribution is given as:

$$f_2(x,\theta) = \frac{\theta^{\left(\frac{n\omega+3n}{2}\right)} \sum x_i^{\omega+2}}{2^{n\left(\frac{\omega+1}{2}\right)} \Gamma\left(\frac{\omega+3}{2}\right)^n} \exp\left(-\frac{\theta}{2}\sum x_{i^2}\right) \frac{\beta^\alpha}{\Gamma(\alpha)} \exp(-\beta\theta)\, \theta^{(\alpha-1)} \tag{22}$$

Also, the Marginal density function of x is given by:

$$f_2(x) = \frac{\sum x_i^{\omega+2}}{2^{n\left(\frac{\omega+1}{2}\right)} \Gamma\left(\frac{\omega+3}{2}\right)^n} \frac{\beta^\alpha}{\Gamma(\alpha)} \frac{\Gamma\left(\frac{n\omega+3n+2\alpha}{2}\right)}{\left(\frac{\sum x_{i^2}}{2} + \beta\right)^{\left(\frac{n\omega+3n+2\alpha}{2}\right)}} \tag{23}$$

Using above Two Results (22) and (23), we get the posterior Probability Density Function:

$$\pi_2(\theta/x) = \frac{\theta^{\frac{n\omega+3n+2\alpha-2}{2}} \left(\frac{\sum x_{i^2}}{2} + \beta\right)^{\left(\frac{n\omega+3n+2\alpha}{2}\right)}}{\Gamma\left(\frac{n\omega+3n+2\alpha}{2}\right)} \exp\left(-\theta\left(\frac{\sum x_{i^2}}{2} + \beta\right)\right) \tag{24}$$

2.2.2.1 Under squared-error loss function

The Risk function under Squared Error Function is given as:

$$R_{(\mathrm{Sq,gp})}(\hat{\theta}) = c\hat{\theta}^2 + \frac{c\left(\frac{n\omega+3n+2\alpha+2}{2}\right)\left(\frac{n\omega+3n+2\alpha}{2}\right)}{\left(\frac{\sum x_{i^2}}{2} + \beta\right)^2} - \frac{2c\hat{\theta}\left(\frac{n\omega+3n+2\alpha}{2}\right)}{\left(\frac{\sum x_{i^2}}{2} + \beta\right)} \tag{25}$$

After solving the above Risk function, we get the Baye's estimator:

$$\hat{\theta}_{(\mathrm{Sq,gp})} = \frac{n\omega + 3n + 2\alpha}{\sum x_{i^2} + 2\beta} \tag{26}$$

2.2.2.2 Under precautionary loss function

The Risk function under precautionary Loss function is given as:

$$R_{(\mathrm{Pr,gp})}(\hat{\theta}) = c\hat{\theta} + c\hat{\theta}^{-1}\frac{\left(\frac{n\omega+3n+2\alpha+2}{2}\right)\left(\frac{n\omega+3n+2\alpha}{2}\right)\Gamma\left(\frac{n\omega+3n+2\alpha}{2}\right)}{\Gamma\left(\frac{n\omega+3n+2\alpha}{2}\right)\left(\frac{\sum x_{i^2}}{2}+\beta\right)^{\left(\frac{4}{2}\right)}} - 2c\frac{\left(\frac{n\omega+3n+2\alpha}{2}\right)}{\left(\frac{\sum x_{i^2}}{2}+\beta\right)^{\left(\frac{2}{2}\right)}} \tag{27}$$

After solving the above Risk function, we get the Baye's estimator:

$$\hat{\theta}_{(\mathrm{Pr,gp})} = \frac{\sqrt{(n\omega+3n+2\alpha+2)(n\omega+3n+2\alpha)}}{\left(\sum x_{i^2}+2\beta\right)} \tag{28}$$

2.2.2.3 Under Al-Bayyati's loss function

The Risk function under Al-Bayyati's Loss function:

$$R_{(\mathrm{Al,gp})}(\hat{\theta}) = \frac{\hat{\theta}^2}{\Gamma\left(\frac{n\omega+3n+2\alpha}{2}\right)}\frac{\Gamma\left(\frac{n\omega+3n+2\alpha+2c_2}{2}\right)}{\left(\frac{\sum x_{i^2}}{2}+\beta\right)^{(c_2)}} + \frac{\Gamma\left(\frac{n\omega+3n+2\alpha+2c_2+4}{2}\right)}{\Gamma\left(\frac{n\omega+3n+2\alpha}{2}\right)\left(\frac{\sum x_{i^2}}{2}+\beta\right)^{(c_2+2)}} - \frac{2\hat{\theta}}{\Gamma\left(\frac{n\omega+3n+2\alpha}{2}\right)}\frac{\Gamma\left(\frac{n\omega+3n+2\alpha+2c_2+2}{2}\right)}{\left(\frac{\sum x_{i^2}}{2}+\beta\right)^{(c_2+1)}} \tag{29}$$

After solving $\frac{\partial R_{(\mathrm{Al,gp})}(\hat{\theta})}{\partial \hat{\theta}} = 0$ for $\hat{\theta}$, we will have the Baye's estimator given by:

$$\hat{\theta}_{\mathrm{Al,gp})} = \frac{n\omega+3n+2\alpha+2c_2}{\sum x_{i^2}+2\beta} \tag{30}$$

2.2.2.4 Under Stein's loss function

The Risk function under Stein Loss Function is given by:

$$R_{(\mathrm{St,gp})}(\hat{\theta}) = \int_0^\infty\left(\frac{\hat{\theta}}{\theta} - \log\frac{\hat{\theta}}{\theta} - 1\right)\frac{\theta^{\frac{n\omega+3n+2\alpha-2}{2}}\left(\frac{\sum x_{i^2}}{2}+\beta\right)^{\left(\frac{n\omega+3n+2\alpha}{2}\right)}}{\Gamma\left(\frac{n\omega+3n+2\alpha}{2}\right)}\exp\left(-\theta\left(\frac{\sum x_{i^2}}{2}+\beta\right)\right)d\theta$$

$$R_{(\mathrm{St,gp})}(\hat{\theta}) = \hat{\theta}\,\frac{\Gamma\left(\frac{n\omega+3n+2\alpha-2}{2}\right)\left(\frac{\sum x_{i^2}}{2}+\beta\right)}{\left(\frac{n\omega+3n+2\alpha-2}{2}\right)\Gamma\left(\frac{n\omega+3n+2\alpha-2}{2}\right)} - \log(\hat{\theta}) + e^{-t} - 1 \tag{31}$$

After solving the above Risk function, we will have required Baye's estimator:

$$\hat{\theta}_{(\mathrm{St,gp})} = \frac{n\omega+3n+2\alpha-2}{\left(\sum x_{i^2}+2\beta\right)} \tag{32}$$

2.2.3 Bayesian estimation under exponential (α) prior distributions

The Joint Probability Density Function of Weighted Maxwell-Boltzmann Distribution Using Exponential Prior Distribution is given as:

$$f_2(x,\theta) = \frac{\theta^{\left(\frac{n\omega+3n}{2}\right)} \sum x_i^{\omega+2}}{2^{n\left(\frac{\omega+1}{2}\right)} \Gamma\left(\frac{\hat{\omega}+3}{2}\right)^n} \exp\left(-\frac{\theta}{2}\sum x_{i^2}\right) \frac{1}{\Gamma(\alpha)} \exp(-\theta)\, \theta^{(\alpha-1)} \tag{33}$$

Also, the Marginal density function of x is given by:

$$f_2(x) = \frac{\sum x_i^{\omega+2}}{2^{n\left(\frac{\omega+1}{2}\right)} \Gamma\left(\frac{\omega+3}{2}\right)^n} \frac{1}{\Gamma(\alpha)} \frac{\Gamma\left(\frac{n\omega+3n+2\alpha}{2}\right)}{\left(\frac{\sum x_{i^2}}{2}+1\right)^{\left(\frac{n\omega+3n+2\alpha}{2}\right)}} \tag{34}$$

Using above Two Results (33) and (34), we get the posterior Probability Density Function:

$$\pi_2(\theta/x) = \frac{\theta^{\frac{n\omega+3n+2\alpha-2}{2}} \left(\frac{\sum x_{i^2}}{2}+1\right)^{\left(\frac{n\omega+3n+2\alpha}{2}\right)}}{\Gamma\left(\frac{n\omega+3n+2\alpha}{2}\right)} \exp\left(-\theta\left(\frac{\sum x_{i^2}}{2}+1\right)\right) \tag{35}$$

2.2.3.1 *Under squared-error loss function*

The Risk function under Squared Error Function is given as:

$$R_{(\text{Sq,Ep})}\left(\hat{\theta}\right) = c\hat{\theta}^2 + \frac{c\left(\frac{n\omega+3n+2\alpha+2}{2}\right)\left(\frac{n\omega+3n+2\alpha}{2}\right)}{\left(\frac{\sum x_{i^2}}{2}+1\right)^2} - \frac{2c\hat{\theta}\left(\frac{n\omega+3n+2\alpha}{2}\right)}{\left(\frac{\sum x_{i^2}}{2}+1\right)} \tag{36}$$

After solving the above Risk function, we get the Baye's estimator:

$$\hat{\theta}_{(\text{Sq,ep})} = \frac{n\omega+3n+2\alpha}{\sum x_{i^2}+2} \tag{37}$$

2.2.3.2 *Under precautionary loss function*

The Risk function under precautionary Loss function is given as:

$$R_{(\text{Pr,ep})}\left(\hat{\theta}\right) = c\hat{\theta} + c\hat{\theta}^{-1} \frac{\left(\frac{n\omega+3n+2\alpha+2}{2}\right)\left(\frac{n\omega+3n+2\alpha}{2}\right)\Gamma\left(\frac{n\omega+3n+2\alpha}{2}\right)}{\Gamma\left(\frac{n\omega+3n+2\alpha}{2}\right)\left(\frac{\sum x_{i^2}}{2}+1\right)^{\left(\frac{4}{2}\right)}} - 2c\frac{\left(\frac{n\omega+3n+2\alpha}{2}\right)}{\left(\frac{\sum x_{i^2}}{2}+1\right)^{\left(\frac{2}{2}\right)}}$$

After solving the above Risk function, we get the Baye's estimator:

$$\hat{\theta}_{(\text{Pr,ep})} = \frac{\sqrt{(n\omega+3n+2\alpha+2)(n\omega+3n+2\alpha)}}{\left(\sum x_{i^2}+2\right)} \tag{38}$$

2.2.3.3 *Under Al-Bayyati's loss function*

The Risk function under Al-Bayyati's Loss function:

$$R_{(\text{Al,ep})}(\hat{\theta}) = \frac{\hat{\theta}^2}{\Gamma\left(\frac{n\omega+3n+2\alpha}{2}\right)} \frac{\Gamma\left(\frac{n\omega+3n+2\alpha+2c_2}{2}\right)}{\left(\frac{\sum x_{i^2}}{2}+1\right)^{(c_2)}} + \frac{\Gamma\left(\frac{n\omega+3n+2\alpha+2c_2+4}{2}\right)}{\Gamma\left(\frac{n\omega+3n+2\alpha}{2}\right)\left(\frac{\sum x_{i^2}}{2}+1\right)^{(c_2+2)}} - \frac{2\hat{\theta}}{\Gamma\left(\frac{n\omega+3n+2\alpha}{2}\right)} \frac{\Gamma\left(\frac{n\omega+3n+2\alpha+2c_2+2}{2}\right)}{\left(\frac{\sum x_{i^2}}{2}+1\right)^{(c_2+1)}} \tag{39}$$

After solving the above Risk Function, we will have the Baye's estimator given by:

$$\hat{\theta}_{\text{Al,Ep})} = \frac{n\omega+3n+2\alpha+2c_2}{\sum x_{i^2}+2} \tag{40}$$

2.2.3.4 Under Stein's loss function

The Risk function under Stein Loss Function is given by:

$$R_{(\text{St,ep})}(\hat{\theta}) = \int_0^\infty \left(\frac{\hat{\theta}}{\theta} - \log\frac{\hat{\theta}}{\theta} - 1\right) \frac{\theta^{\frac{n\omega+3n+2\alpha-2}{2}}\left(\frac{\sum x_{i^2}}{2}+1\right)^{\left(\frac{n\omega+3n+2\alpha}{2}\right)}}{\Gamma\left(\frac{n\omega+3n+2\alpha}{2}\right)} \exp\left(-\theta\left(\frac{\sum x_{i^2}}{2}+1\right)\right)d\theta$$

$$R_{(\text{St,ep})}(\hat{\theta}) = \hat{\theta}\,\frac{\Gamma\left(\frac{n\omega+3n+2\alpha-2}{2}\right)\left(\frac{\sum x_{i^2}}{2}+1\right)}{\left(\frac{n\omega+3n+2\alpha-2}{2}\right)\Gamma\left(\frac{n\omega+3n+2\alpha-2}{2}\right)} - \log\left(\hat{\theta}\right) + e^{-t} - 1 \tag{41}$$

After solving the above Risk function, we will have required Baye's estimator:

$$\hat{\theta}_{(\text{St,ep})} = \frac{n\omega+3n+2\alpha-2}{\left(\sum x_{i^2}+2\right)} \tag{42}$$

3. Simulation study of weighted Maxwell-Boltzmann distribution

In this section, we conduct the simulation studies of weighted Maxwell-Boltzmann distribution to examine the performance of the MLEs and Bayesian estimators under different prior's like extension of Jeffrey's' prior, Gamma prior and Exponential prior under different loss functions in terms of expected estimates, biases, variances and mean squared errors by considering different parameter combinations. For the simulation study, sample size is taken as n = (25, 100, 300) to observe the effect of small, moderate and large samples on the estimators. We have conducted the simulation procedure for different random parameter combinations and the process was repeated 2000 times. From the simulation results, it is concluded that the performances of the Bayesian and MLEs become better when the sample size increases. In terms of MSE, the Bayesian estimators under Gamma prior perform better (see **Table 1**). In specific, from **Table 2**, extension of Jeffery's prior under Al-Bayyati's error loss function and stein's loss function gives smaller MSE's as compared to other loss functions.

From **Table 1**, we can see that the performances of the Bayesian and MLEs become better when the sample size increases. For large samples, Gamma prior

n	θ	ω	α	β	c_2	**Criterion**	$\hat{\theta}_{ml}$	$\hat{\theta}_{sq}$	θ_{pre}	$\hat{\theta}_{alb}$	θ_{ste}
25	0.5	0.5	0.5	0.5	0.5	$E(\theta)$	0.504729	0.527842	0.526203	0.511548	0.532156
						Bias	0.004729	0.027842	0.026203	0.011548	0.032156
						Variance	0.003753	0.007177	0.006050	0.003911	0.007792
						MSE	0.003775	0.007952	0.006737	0.004044	0.008826
100	0.5	0.5	0.5	0.5	0.5	$E(\theta)$	0.506417	0.501861	0.502694	0.504235	0.500950
						Bias	0.006417	0.001861	0.002694	0.004235	0.000950
						Variance	0.001238	0.001789	0.001571	0.001552	0.001159
						MSE	0.001279	0.001792	0.001578	0.001570	0.001160
300	0.5	0.5	0.5	0.5	0.5	$E(\theta)$	0.500341	0.501755	0.500753	0.506296	0.501444
						Bias	0.000341	0.001755	0.000753	0.006296	0.001444
						Variance	0.000434	0.000593	0.000357	0.000464	0.000416
						MSE	0.000434	0.000596	0.000358	0.000503	0.000418
25	1.5	0.5	0.4	0.5	0.5	$E(\theta)$	1.555484	1.497774	1.547039	1.552606	1.478335
						Bias	0.055484	−0.002226	0.047039	0.052606	−0.021665
						Variance	0.059091	0.055145	0.057945	0.054047	0.042449
						MSE	0.062169	0.055150	0.060158	0.056815	0.042918
100	1.5	0.5	0.4	0.5	0.5	$E(\theta)$	1.511161	1.487596	1.479585	1.506022	1.510950
						Bias	0.011161	−0.012404	−0.020415	0.006022	0.010950
						Variance	0.010559	0.012894	0.013387	0.012110	0.011800
						MSE	0.010683	0.013048	0.013804	0.012146	0.011919

n	θ	ω	α	β	c_2	Criterion	$\hat{\theta}_{ml}$	$\hat{\theta}_{sq}$	θ_{pre}	$\hat{\theta}_{alb}$	θ_{ste}
300	1.5	0.5	0.4	0.5	0.5	$E(\theta)$	1.500168	1.510651	1.493056	1.514740	1.502537
						Bias	0.000168	0.010651	−0.006944	0.014740	0.002537
						Variance	0.005194	0.003724	0.004315	0.005630	0.004500
						MSE	0.005194	0.003838	0.004363	0.005847	0.004506
25	2.0	1.6	0.4	1.5	1.5	$E(\theta)$	1.938975	1.991177	2.000771	1.927722	1.956042
						Bias	−0.061025	−0.008823	0.000771	−0.072278	−0.043958
						Variance	0.059838	0.070756	0.103288	0.065984	0.061025
						MSE	0.063562	0.070833	0.103288	0.071208	0.062957
100	2.0	1.6	0.4	1.5	1.5	$E(\theta)$	1.969832	1.963989	1.977604	2.015424	1.975888
						Bias	−0.030168	−0.036011	−0.022396	0.015424	−0.024112
						Variance	0.016741	0.017403	0.012035	0.016688	0.014520
						MSE	0.017651	0.018700	0.012537	0.016926	0.015102
300	2.0	1.6	0.4	1.5	1.5	$E(\theta)$	2.004078	1.987208	1.994396	2.003804	2.002971
						Bias	0.004078	−0.012792	−0.005604	0.003804	0.002971
						Variance	0.006379	0.006670	0.005228	0.005783	0.008284
						MSE	0.006396	0.006834	0.005260	0.005798	0.008293
25	2.5	1.6	0.4	1.5	2.0	$E(\theta)$	2.394049	2.404621	2.347984	2.500847	2.458514
						Bias	−0.105951	−0.095379	−0.152016	0.000847	−0.041486
						Variance	0.098064	0.118924	0.119786	0.154811	0.094105
						MSE	0.109290	0.128021	0.142895	0.154812	0.095827

n	θ	ω	α	β	c_2	Criterion	$\hat{\theta}_{ml}$	$\hat{\theta}_{sq}$	θ_{pre}	$\hat{\theta}_{alb}$	θ_{ste}
100	2.5	1.6	0.4	1.5	2.0	$E(\theta)$	2.515576	2.494660	2.491992	2.484427	2.508391
						Bias	0.015576	−0.005340	−0.008008	−0.015573	0.008391
						Variance	0.022966	0.026813	0.021573	0.023537	0.023719
						MSE	0.023209	0.026842	0.021637	0.023780	0.023790
300	2.5	1.6	0.4	1.5	2.0	$E(\theta)$	2.505136	2.490348	2.502325	2.488103	2.519407
						Bias	0.005136	−0.009652	0.002325	−0.011897	0.019407
						Variance	0.008742	0.008224	0.006055	0.010543	0.009009
						MSE	0.008768	0.008317	0.006060	0.010684	0.009386

ml, *maximum likelihood;* sq, *squared error loss function;* pre, *precautionary loss function;* alb, *Al-Bayyati's loss function;* ste, *Stein's loss function.*

Table 1.
Average estimate, bias, variance and mean squared error for $(\hat{\theta})$ *under gamma prior.*

n	θ	ω	c_1	c_2	Criterion	$\hat{\theta}_{ml}$	$\hat{\theta}_{sq}$	θ_{pre}	$\hat{\theta}_{alb}$	θ_{ste}
25	0.5	0.5	0.5	0.5	$E(\theta)$	0.504550	0.508841	0.519292	0.520839	0.525694
					Bias	0.004550	0.008841	0.019292	0.020839	0.025694
					Variance	0.004979	0.005660	0.008970	0.006678	0.009465
					MSE	0.004999	0.005738	0.009342	0.007112	0.010125
100	0.5	0.5	0.5	0.5	$E(\theta)$	0.505610	0.508184	0.501928	0.492615	0.501255
					Bias	0.005610	0.008184	0.001928	−0.007385	0.001255
					Variance	0.001313	0.001677	0.001707	0.001318	0.001377
					MSE	0.001344	0.001744	0.001710	0.001372	0.001379
300	0.5	0.5	0.5	0.5	$E(\theta)$	0.495559	0.503048	0.503241	0.500495	0.504760
					Bias	−0.004441	0.003048	0.003241	0.000495	0.004760
					Variance	0.000391	0.000500	0.000467	0.000532	0.000487
					MSE	0.000411	0.000510	0.000478	0.000533	0.000510
25	1.6	0.5	0.5	0.5	$E(\theta)$	1.656361	1.652441	1.590889	1.596342	1.624334
					Bias	0.056361	0.052441	−0.009111	−0.003658	0.024334
					Variance	0.072186	0.080721	0.048146	0.046070	0.052856
					MSE	0.075363	0.083471	0.048229	0.046084	0.053448
100	1.6	0.5	0.5	0.5	$E(\theta)$	1.584041	1.592563	1.627598	1.617623	1.621388
					Bias	−0.015959	−0.007437	0.027598	0.017623	0.021388
					Variance	0.013766	0.013701	0.012200	0.011394	0.014179
					MSE	0.014021	0.013756	0.012961	0.011704	0.014636
300	1.6	0.5	0.5	0.5	$E(\theta)$	1.595942	1.621590	1.615841	1.605064	1.600224
					Bias	−0.004058	0.021590	0.015841	0.005064	0.000224
					Variance	0.006184	0.006401	0.005696	0.003994	0.005462
					MSE	0.006200	0.006867	0.005947	0.004020	0.005462
25	2.5	1.0	1.5	0.5	$E(\theta)$	2.479248	2.477210	2.416603	2.493049	2.428281
					Bias	−0.020752	−0.022790	−0.083397	−0.006951	−0.071719
					Variance	0.128038	0.127245	0.113067	0.149621	0.111466
					MSE	0.128468	0.127765	0.120022	0.149670	0.116610
100	2.5	1.0	1.5	0.5	$E(\theta)$	2.498178	2.502858	2.527984	2.482989	2.488065
					Bias	−0.001822	0.002858	0.027984	−0.017011	−0.011935
					Variance	0.037896	0.029747	0.030859	0.020501	0.023293
					MSE	0.037899	0.029756	0.031642	0.020791	0.023436
300	2.5	1.0	1.5	0.5	$E(\theta)$	2.510585	2.500611	2.490367	2.507471	2.477393
					Bias	0.010585	0.000611	−0.009633	0.007471	−0.022607
					Variance	0.010037	0.009817	0.007991	0.008569	0.012196
					MSE	0.010149	0.009818	0.008083	0.008625	0.012707
25	2.5	1.0	0.5	1.5	$E(\theta)$	2.566472	2.690959	2.607021	2.559121	2.582644
					Bias	0.066472	0.190959	0.107021	0.059121	0.082644
					Variance	0.133694	0.132587	0.132802	0.133658	0.156276
					MSE	0.138112	0.169052	0.144256	0.137153	0.163106

n	θ	ω	c_1	c_2	Criterion	$\hat{\theta}_{\text{ml}}$	$\hat{\theta}_{\text{sq}}$	θ_{pre}	$\hat{\theta}_{\text{alb}}$	θ_{ste}
100	2.5	1.0	0.5	1.5	$E(\theta)$	2.505612	2.535815	2.524166	2.489736	2.514349
					Bias	0.005612	0.035815	0.024166	−0.010264	0.014349
					Variance	0.042607	0.029863	0.029906	0.031724	0.032759
					MSE	0.042639	0.031146	0.030490	0.031829	0.032965
300	2.5	1.0	0.5	1.5	$E(\theta)$	2.499279	2.508299	2.510490	2.490229	2.487861
					Bias	−0.000721	0.008299	0.010490	−0.009771	−0.012139
					Variance	0.011396	0.009322	0.011898	0.011426	0.011835
					MSE	0.011397	0.009391	0.012008	0.011522	0.011982

ml, *maximum likelihood;* sq, *squared error loss function;* pre, *precautionary loss function;* alb, *Al-Bayyati's loss function;* ste, *Stein's loss function.*

Table 2.
Average estimate, bias, variance and mean squared error for $(\hat{\theta})$ *under extension of Jeffery's prior.*

under squared error loss function and Al-Bayyati's loss function gives smaller MSE's as compared to other loss functions and MLEs.

From **Table 3**, we can see that the performances of the Bayesian and MLEs become better when the sample size increases. Exponential prior under squared error loss function and stein's loss function gives smaller MSE's as compared to other loss functions. Thus, Exponential prior under squared error loss function and stein's loss function can be preferred for parameter estimation.

4. Applications of weighted Maxwell-Boltzmann distribution

In this section, we present the goodness of fit of weighted Maxwell-Boltzmann distribution (WMB). For testing the goodness of fit of weighted Maxwell-Boltzmann distribution over Maxwell-Boltzmann (MB), length biased Maxwell-Boltzmann (LBMB) and area biased Maxwell-Boltzmann (ABMB) distributions, following two data sets have been considered.

Data set I is regarding tensile strength, measured in GPA, of 69 carbon fibers tested under tension at gauge lengths of 20 mm, Bader and Priest [29].

From **Table 4**, it has been observed that weighted Maxwell-Boltzmann distribution have the lesser AIC, AICC, $-\log L$ and BIC values as compared to Maxwell-Boltzmann, length biased Maxwell-Boltzmann and area biased Maxwell-Boltzmann distributions. Hence we can conclude that the Weighted Maxwell-Boltzmann distribution leads to a better fit than the Maxwell-Boltzmann, length biased Maxwell-Boltzmann and area biased Maxwell-Boltzmann distributions in case of analyzing the data set I.

Data set II is regarding the strength data and it represents the strength measured in GPA for single carbon fibers and impregnated 1000-carbon fiber tows. Single fibers were tested under tension at gauge lengths of 10 mm with sample sizes n = 63; see Bader and Priest [29] and Surles and Padgett [30].

From **Table 5**, it has been observed that weighted Maxwell-Boltzmann distribution have the lesser AIC, AICC, $-\log L$ and BIC values as compared to Maxwell-Boltzmann, length biased Maxwell-Boltzmann and area biased Maxwell-Boltzmann distributions. Hence we can conclude that the Weighted Maxwell-Boltzmann distribution leads to a better fit than the Maxwell-Boltzmann, length biased Maxwell-Boltzmann and area biased Maxwell-Boltzmann distributions in case of analyzing the data set II.

n	θ	ω	α	c_2	Criterion	$\hat{\theta}_{ml}$	$\hat{\theta}_{sq}$	θ_{pre}	$\hat{\theta}_{alb}$	θ_{ste}
25	0.5	0.5	0.5	0.5	$E(\theta)$	0.489954	0.518542	0.508290	0.497323	0.494616
					Bias	−0.010046	0.018542	0.008290	−0.002677	−0.005384
					Variance	0.006497	0.005296	0.005568	0.005993	0.005576
					MSE	0.006598	0.005640	0.005636	0.006000	0.005605
100	0.5	0.5	0.5	0.5	$E(\theta)$	0.501216	0.510728	0.499014	0.503758	0.508104
					Bias	0.001216	0.010728	−0.000986	0.003758	0.008104
					Variance	0.000940	0.001242	0.001713	0.001204	0.001588
					MSE	0.000941	0.001357	0.001714	0.001218	0.001654
300	0.5	0.5	0.5	0.5	$E(\theta)$	0.500133	0.496843	0.497977	0.499664	0.499775
					Bias	0.000133	−0.003157	−0.002023	−0.000336	−0.000225
					Variance	0.000413	0.000407	0.000447	0.000357	0.000340
					MSE	0.000413	0.000417	0.000451	0.000357	0.000340
25	1.5	0.5	0.4	0.5	$E(\theta)$	1.483379	1.482564	1.511903	1.540451	1.566552
					Bias	−0.016621	−0.017436	0.011903	0.040451	0.066552
					Variance	0.066483	0.048448	0.049620	0.060298	0.047485
					MSE	0.066759	0.048752	0.049761	0.061934	0.051914
100	1.5	0.5	0.4	0.5	$E(\theta)$	1.490401	1.508552	1.501250	1.501611	1.472398
					Bias	−0.009599	0.008552	0.001250	0.001611	−0.027602
					Variance	0.013307	0.012242	0.014316	0.015729	0.012469
					MSE	0.013399	0.012316	0.014317	0.015731	0.013231

n	θ	ω	α	c_2	Criterion	$\hat{\theta}_{ml}$	$\hat{\theta}_{sq}$	θ_{pre}	$\hat{\theta}_{alb}$	θ_{ste}
300	1.5	0.5	0.4	0.5	$E(\theta)$	1.517577	1.511777	1.521694	1.499300	1.505245
					Bias	0.017577	0.011777	0.021694	−0.000700	0.005245
					Variance	0.004328	0.004223	0.003876	0.003398	0.003845
					MSE	0.004637	0.004362	0.004346	0.003398	0.003873
25	2.0	1.6	0.4	1.5	$E(\theta)$	2.044911	2.009608	1.985619	1.980819	1.993491
					Bias	0.044911	0.009608	−0.014381	−0.019181	−0.006509
					Variance	0.112842	0.062620	0.057835	0.085826	0.071638
					MSE	0.114859	0.062712	0.058042	0.086194	0.071680
100	2.0	1.6	0.4	1.5	$E(\theta)$	1.996708	1.997635	2.008162	2.021764	2.002779
					Bias	−0.003292	−0.002365	0.008162	0.021764	0.002779
					Variance	0.015238	0.012785	0.013466	0.014708	0.015945
					MSE	0.015248	0.012791	0.013533	0.015181	0.015953
300	2.0	1.6	0.4	1.5	$E(\theta)$	1.997046	2.002154	1.992272	2.005303	2.004678
					Bias	−0.002954	0.002154	−0.007728	0.005303	0.004678
					Variance	0.006380	0.005365	0.005987	0.006364	0.006716
					MSE	0.006389	0.005370	0.006046	0.006392	0.006738
25	2.5	1.6	0.4	2.0	$E(\theta)$	2.471114	2.476397	2.496794	2.507617	2.539850
					Bias	−0.028886	−0.023603	−0.003206	0.007617	0.039850
					Variance	0.105051	0.096227	0.163694	0.108075	0.135941
					MSE	0.105885	0.096785	0.163704	0.108134	0.137529

n	θ	ω	α	c_2	Criterion	$\hat{\theta}_{ml}$	$\hat{\theta}_{sq}$	θ_{pre}	$\hat{\theta}_{alb}$	θ_{ste}
100	2.5	1.6	0.4	2.0	$E(\theta)$	2.507110	2.499740	2.498220	2.517427	2.508359
					Bias	0.007110	−0.000260	−0.001780	0.017427	0.008359
					Variance	0.024000	0.027471	0.033577	0.031666	0.028237
					MSE	0.024051	0.027471	0.033581	0.031970	0.028307
300	2.5	1.6	0.4	2.0	$E(\theta)$	2.515885	2.503115	2.476479	2.507093	2.504668
					Bias	0.015885	0.003115	−0.023521	0.007093	0.004668
					Variance	0.011382	0.007855	0.011195	0.009851	0.007705
					MSE	0.011635	0.007865	0.011749	0.009901	0.007727

ml, *maximum likelihood;* sq, *squared error loss function;* pre, *precautionary loss function;* alb, *Al-Bayyati's loss function;* ste, *Stein's loss function.*

Table 3.
Average estimate, bias, variance and mean squared error for $(\hat{\theta})$ *under exponential prior.*

Distribution	α_{ml}	θ_{ml}	$-2\log(l)$	AIC	BIC	AICC
WMB	9.079	1.923	50.393	104.787	109.255	104.968
MB	0	0.478	74.633	151.265	153.499	151.325
LBMB	1	0.637	66.713	135.426	137.660	135.485
ABMB	2	0.796	61.385	124.770	127.004	124.829

Table 4.
*Model comparison using AIC, AICC, BIC and -log*L *criterion for data set 1.*

Distribution	α_{ml}	θ_{ml}	$-2\log(l)$	AIC	BIC	AICC
WMB	9.971	1.332	57.656	119.311	123.598	119.511
MB	0	0.308	81.585	165.170	167.313	165.235
LBMB	1	0.411	74.165	150.330	152.473	150.395
ABMB	2	0.513	69.111	140.222	142.366	140.288

Table 5.
*Model comparison using AIC, AICC, BIC and -log*L *criterion for data set II.*

5. Conclusions

1. From the simulation Study, it was observed that the performances of the Bayesian and MLEs become better, when the sample size increases.

2. In terms of MSE, the Bayesian estimators under Gamma prior perform better. In specific, extension of Jeffery's prior under Al-Bayyati's error loss function and stein's loss function gives smaller MSE's as compared to other loss functions.

3. For large samples, Gamma prior under squared error loss function and Al-Bayyati's loss function gives smaller MSE's as compared to other loss functions and MLEs. Exponential prior under squared error loss function and stein's loss function gives smaller MSE's as compared to other loss functions.

4. Thus, Exponential prior under squared error loss function and stein's loss function can be preferred for parameter estimation.

5. It has been observed that weighted Maxwell-Boltzmann distribution have the lesser AIC, AICC, $-\log L$ and BIC values as compared to Maxwell-Boltzmann, length biased Maxwell-Boltzmann and area biased Maxwell-Boltzmann distributions. Hence we can conclude that the Weighted Maxwell-Boltzmann distribution leads to a better fit than the Maxwell-Boltzmann, length biased Maxwell-Boltzmann and area biased Maxwell-Boltzmann distributions in case of analyzing the data set I and II.

Acknowledgements

The authors are highly thankful to the referees and the editor for their valuable suggestions.

Author details

Javaid Ahmad Reshi[1]*, Bilal Ahmad Para[2] and Shahzad Ahmad Bhat[3]

1 Department of Statistics, Government Degree College, Anantnag, India

2 Department of Mathematical Sciences, Islamic University of Science and Technology, Awantipora, India

3 Department of Commerce, Government Degree College, Anantnag, India

*Address all correspondence to: reshijavaid19@gmail.com

References

[1] Tyagi, R. K., and Bhattacharya, S. K. (1989). Bayes estimation of the Maxwell's velocity distribution function. Statistica 29(4), 563–567.

[2] Chaturvedi, A., and Rani, U. (1998). Classical and Bayesian reliability estimation of the generalized Maxwell failure distribution. Journal of Statistical Research 32, 113–120.

[3] Bekker, A. J. J. J., and Roux, J. J. J. (2005). Reliability characteristics of the Maxwell distribution: A Bayes estimation study. Communications in Statistics-Theory and Methods 34(11), 2169–2178.

[4] Kazmi, S., Aslam, M., and Ali, S. (2012). On the Bayesian estimation for two component mixture of Maxwell distribution, assuming type I censored data. International Journal of Applied Science and Technology 2(1), 197–218.

[5] Maxwell, J. C. (1990). The Scientific Letters and Papers of James Clerk Maxwell. Cambridge: Cambridge University Press. pp. 1846–1862.

[6] Aijaz, A., Ahmed, A., Reshi. J. A. (2017). Bayesian analysis of Maxwell-Boltzmann distribution under different loss functions and prior distributions. Pakistan Journal of Statistics 33(6), 419–440.

[7] Huang, J., and Chen, S. (2016). Tail behavior of the generalized Maxwell distribution. Communications in Statistics-Theory and Methods 45(14), 4230–4236.

[8] Krishna, H., and Malik, M. (2012). Reliability estimation in Maxwell distribution with progressively type-II censored data. Journal of Statistical Computation and Simulation 82(4), 623–641.

[9] Tomer, S. K., and Panwar, M. S. (2015). Estimation procedures for Maxwell distribution under type-I progressive hybrid censoring scheme. Journal of Statistical Computation and Simulation 85(2), 339–356.

[10] Zhang, W., Guo, X., and Kassab, G. S. (2008). A generalized Maxwell model for creep behavior of artery opening angle. Journal of Biomechanical Engineering 130(5), 1–16.

[11] Monsia, M. D. (2011). A simplified nonlinear generalized Maxwell model for predicting the time dependent behavior of viscoelastic materials. World Journal of Mechanics 1, 158–167.

[12] Al-Baldawi, T. H. (2013). Comparison of maximum likelihood and some Bayes estimators for Maxwell distribution based on non-informative priors. Baghdad Science Journal 10(2), 480–488.

[13] Dey, S., Dey, T., and Maiti, S. S. (2013). Bayesian inference for Maxwell distribution under conjugate prior. Model Assisted Statistics and Applications 8(3), 193–203.

[14] Podder, C. K., and Roy, M. K. (2003). Bayesian estimation of the parameter of Maxwell distribution under MLINEX loss function. Journal of Statistical Studies 23, 11–16.

[15] Rasheed, H. A. (2013). Minimax estimation of the parameter of the Maxwell distribution under quadratic loss function. Journal of Al-Rafidain University College 31, 43–56.

[16] Spiring, F.A., and Yeung, A. S. (1998). A general class of loss functions with industrial applications. Journal of Quality Technology 30(2), 152–162.

[17] Fisher, R. A. (1934). The effects of methods of ascertainment upon the estimation of frequencies. Ann. Eugenics 6, 13–25.

[18] Rao, C. R. (1965). On discrete distributions arising out of method of ascertainment. In: Classical and Contagious Discrete. G. P. Patil, editor. Calcutta: Pergamon Press and Statistical Publishing Society. pp. 320–332.

[19] Dar, A. A., Ahmed, A., and Reshi, J. A. (2018). Characterisation and estimation of weighted Maxwell-Boltzmann distribution. Applied Mathematics and Information Sciences 12(1), 193–202.

[20] Aslam, M. (2003). An application of prior predictive distribution to elicit the prior density. Journal of Statistical Theory and Applications 2(1), 70–83.

[21] K. Ahmad, S. P. Ahmad, A. Ahmed, J. A. Reshi (2015). Bayesian analysis of generalized gamma distribution using R software. Journal of Statistics Applications and Probability 4(2), 323–335.

[22] Norstrom, J. G. (1996). The use of precautionary loss functions in risk analysis. IEEE Transactions on Reliability 45(3), 400–403.

[23] Ahmed, A., Ahmad, S. P., and Reshi, J. A. (2013). Bayesian analysis of Rayleigh distribution. International Journal of Scientific and Research Publications 3(10), 217–225.

[24] Zellner, A. (1986). Bayesian estimation and prediction using asymmetric loss functions. Journal of the American Statistical Association 81 (394), 446–451.

[25] Reshi, J. A., Ahmed, A., and Mir, K. A. (2014). Bayesian estimation of size biased classical gamma distribution. Journal of Reliability and Statistical Studies 7(1), 31–42.

[26] Dey, S., and Maiti, S. S. (2010). Bayesian estimation of the parameter of Maxwell distribution under different loss functions. Journal of Statistical Theory and Practice 4(2), 279–287.

[27] Al-Kutubi, H. S. (2005). On comparison estimation procedures for parameter and survival function exponential distribution using simulation [Ph.D. thesis]. Iraq: Baghdad University.

[28] Wald, A. (1950). Statistical Decision Functions. New York: John Wiley and Sons.

[29] Bader, M. G., and Priest, A. M. (1982). Statistical aspects of fiber and bundle strength in hybrid composites. In: Hayashi, T., Kawata, K., Umekawa, S., editors. Progress in Science in Engineering Composites. Tokyo: ICCM-IV. pp. 1129–1136.

[30] Surles, J. G., and Padgett, W. J. (2001). Inference for reliability and stress-strength for a scaled Burr type X distribution. Lifetime Data Anal 7, 187–200.

Chapter 6

Averaging Indoor Localization System

Eman Shawky Abd El-Fattah Amer

Abstract

This chapter aims at improving the accuracy of estimation the localization by using the RSS method to estimate the positions and take into account the effects of both LOS propagation. The proposed system is depending on developing a mathematical model for the noisy VLC positioning system. For improving the results, adopting the KF is combined with the proposed system, which is considered an optimal estimator. The performance of the proposed technique is determined by evaluating the positioning errors in a typical room. Also this chapter develops the accuracy of the positioning system by using different ideas with average techniques. The discussion of the results for averaging technique is displayed.

Keywords: RSS, KF, VLC, localization

1. Introduction

This chapter aims at improving the accuracy of estimation the localization by using the Received Signal Strength (RSS) method to estimate the positions and take into account the effects of both line of sight (LOS) and non-line of sight (NLoS) propagations. The proposed system is depending on developing a mathematical model for the noisy Visible light communication (VLC) positioning system. For improving the results, adopting the Kalman filter (KF) is combined with the proposed system, which is considered an optimal estimator. The performance of the proposed technique is determined by evaluating the positioning errors in a typical room. Also, this chapter develops the accuracy of the positioning system by using different ideas with average techniques.

The remaining of this chapter is organized as follows: Section 2 discusses the optical channel in indoor systems. Section 3 is devoted to explaining the methodology of localization using RSS techniques. A mathematical derivation for performance evaluation is developed in the same section as well. In Section 4, The proposed KF algorithm is presented with explaining its algorithm for estimation correction. There is an average technique aiming to use the average method as shown in Section 5. Using both effects of LOS and the first reflection of NLoS propagation is done in the average proposed system. Adopting KF with averaging is shown in Section 6. The discussion of the results for averaging technique is displayed in Section 7. Section 8 shows the comparison between the results with some recent references. Finally, the concluding remarks are given in Section 9.

2. Optical channel model

The characteristics of the channel modeling have been analyzed with the effects of the channel distortions in [1]. The power associated with the channel is separated into two factors, these being optical path loss (PL) and multipath dispersion. The PL is calculated from the knowledge of the receiver size, the transmitter beam divergence, and separation distance. However, a NLoS configuration (diffuse systems) mainly used in the indoor environment, uses reflections of the room surfaces and furniture. These reflections could be seen as unwanted signals or multi-path distortions which predict the PL more complex. The OW channel transfer function is defined by

$$H = H_{LoS} + H_{NLoS} \tag{1}$$

According to **Figure 1**, it describes H_{loS} as the contribution due to the LoS, which is independent of the modulation frequency and it depends on the distance between transmitter and receiver. In a VLC system, the direct current (DC) gain of a VLC channel is expressed by.

$$H^{i}_{\text{LOS}} = \frac{m+1}{2\pi d_i^2} \cos^m(\phi_i) A_R \cos(\psi_i) T_s(\psi_i) g(\psi_i), \tag{2}$$

The received power therefore becomes

$$P_r = H_{LOS}.P_t, \tag{3}$$

where P_t is a transmitted power m is the Lambertian order, d_i is the distance between transmitter i and the receiver, ϕ_i is the irradiance angle, ψ_i is the incidence angle, $T_s(\cdot)$ and $g(\cdot)$ are the gains of the optical filter and concentrator at the receiver (assumed here as unity gain), and A_R is the detector effective area. Where the channel DC gain H_{NLOS} of the first reflection is shown as in the following equation related to **Figure 1**.

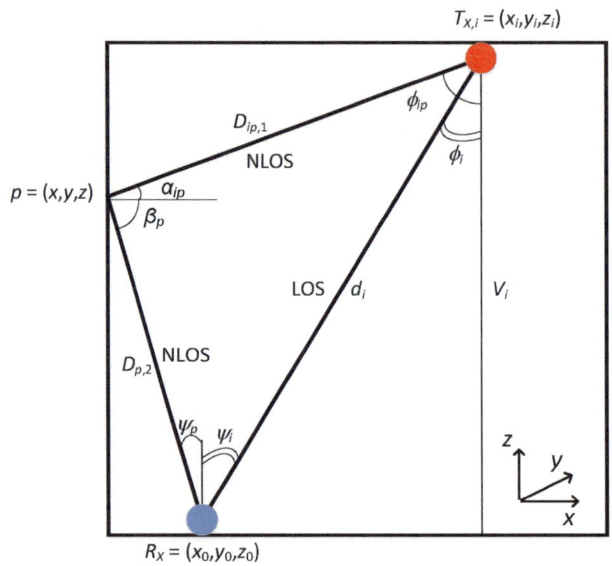

Figure 1.
The channel model of VLC system.

$$\begin{aligned} dH_{\text{NLOS}}^{ip} &= \frac{m+1}{2\pi D_{ip,1}^2 D_{p,2}^2} \cos^m\left(\phi_{ip}\right) \cos\left(\alpha_{ip}\right) \cdot dA_p \\ &\quad \times \rho \cos\left(\beta_p\right) \cos\left(\psi_p\right) T_s\left(\psi_p\right) g\left(\psi_p\right) A_R, \end{aligned} \tag{4}$$

where $D_{ip,1}$ is the distance between transmitter i and reflection point p, $D_{p,2}$ is the distance between reflection point p and receiver R_X, ϕ_{ip} and ψ_p are the NLoS irradiance and incidence angles with respect to point p, respectively, α_{ip} and β_p are the incidence and irradiance angles at reflection point p on the wall, respectively, ρ is the wall reflectivity, and dA_p represents the area of the reflection point on the wall.

The total NLoS channel gain for i th transmitter H_{NLOS}^i is given by collecting the reflections from the 4 walls [2]:

$$H_{\text{NLOS}}^i = \sum_{j=1}^{4} H_{\text{NLOS},wallj}^i, \tag{5}$$

where $H_{\text{NLOS},wallj}^i$ is the collection of reflections from transmitter i to wall j, and can be obtained by integrating (4) over (x,z) or (y,z) based on the wall location, such that

$$\begin{aligned} H_{\text{NLOS},wallj}^i &= \iint\limits_{\substack{(x,z)\text{or} \\ (y,z)}} \frac{m+1}{2\pi D_{ip,1}^2 D_{p,2}^2} \cos^m\left(\phi_{ip}\right) \cos\left(\alpha_{ip}\right)\rho \\ &\quad \times \cos\left(\beta_p\right) \cos\left(\psi_p\right) T_s\left(\psi_p\right) g\left(\psi_p\right) A_R dA_p. \end{aligned} \tag{6}$$

The determination of the parameters for the previous equation is as follows:

$$\begin{aligned} D_{ip,1} &= \sqrt{(T_{X,i} - p)(T_{X,i} - p)^T} \\ D_{p,2} &= \sqrt{(p - R_X)(p - R_X)^T}, \end{aligned} \tag{7}$$

where $()^T$ is the transpose operator for row vector a. By using triangle calculations, the angles $\phi_{ip}, \alpha_{ip}, \psi_p$, and β_p can be found as follows:

$$\begin{aligned} \cos\left(\phi_{ip}\right) &= \frac{|z_i - z|}{D_{ip,1}}, & \alpha_{ip} &= \frac{\pi}{2} - \phi_{ip}, \\ \cos\left(\psi_p\right) &= \frac{|z - z_0|}{D_{p,2}}, & \beta_p &= \frac{\pi}{2} - \psi_p. \end{aligned} \tag{8}$$

The simulation of the power distribution is done for two cases, first case is using four transmitters which are distributed in different positions (1.25, 1.25, 3) m, (1.25, 3.75,3) m, (3.75, 1.25, 3) m, (3.75, 3.75, 3) m in room with size (5,5,3) m, as shown in **Figure 2**. Second case is using only one transmitter that is located in the center of the ceiling. This power distribution is shown in **Figure 3**. The simulation is done with using some parameters where $FOV = 70^o$ and assume the filter gain = 1, the number of LEDs by array 60x60. The concentrator gain = 1 where the active area of photo diode (PD) = $1cm^2$.

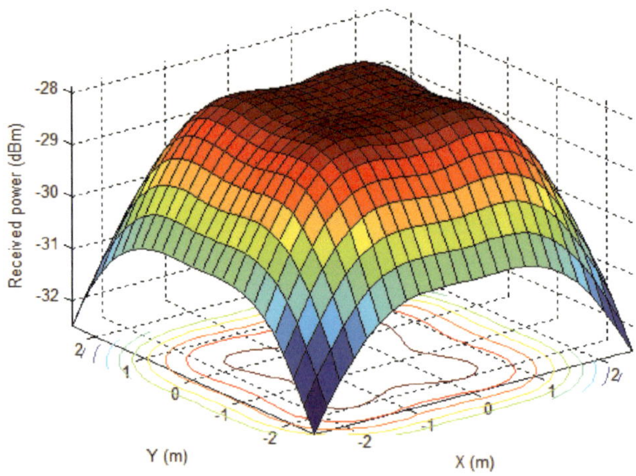

Figure 2.
The power distribution for 4 LEDs.

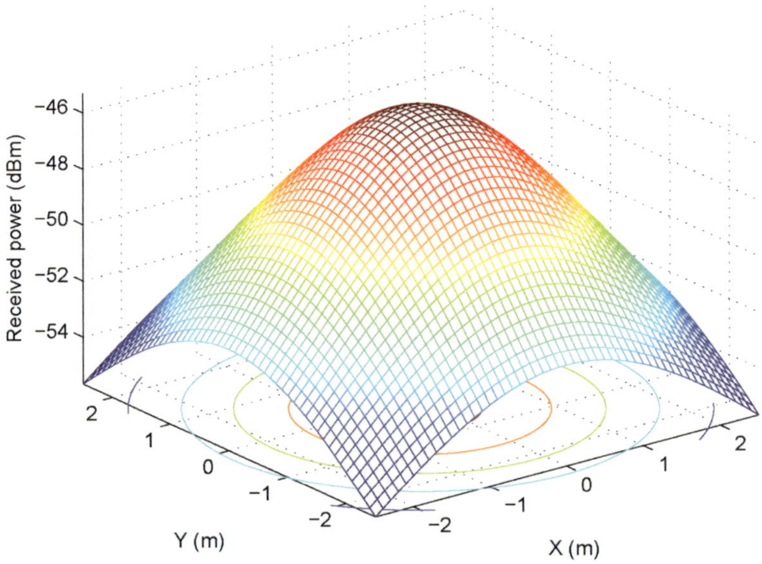

Figure 3.
The power distribution for one LED centered in the ceiling.

3. RSS mathematical analysis

The proposed technique depends on estimating the receiver position using RSS method, then further improving the acquired estimation by adopting the Kalman filtering algorithm. In the initial estimation, RSS technique is used taking into account the effect of LoS. Specifically, A mathematical model is developed for the noisy VLC positioning system and estimate both the angular and horizontal-distance errors. Because of the dependence of horizontal-distance error on the irradiance angle error. The performance of the proposed technique is determined by evaluating the positioning errors in a typical room. Also, the results are compared to that of the traditional RSS system. Depending on **Figure 1**; the analysis assumes that

$\psi_i = \phi_i$ for any $i \in \{1, 2, 3, 4\}$. Assume that the light emitted from each LED is distinguishable at the receiver. Accordingly, The index i is dropped from the developed equations.

If the transmitter and receiver are aligned together, then $\psi = \phi = 0, d = V$, where d is a direct distance between transmitter and receiver and V is the horizontal distance of the transmitter, using $T_s(\psi_i) = 1$, and $g(\psi_i) = 1$ as shown in **Figure 1**. In this case, the power of LoS can be approximated as the following:

$$P_{RO} = \frac{(m+1)A_R}{2\pi V^2} P_T. \tag{9}$$

In the general case ($\phi \neq 0$) as view in (2), and by using **Figure 1** where $\cos^2(\phi) = \frac{V^2}{d^2}$ and by multiplying the equation by $\frac{V^2}{V^2}$, then, the received power can be modeled as:

$$P_R = \cos^{m+3}(\phi) P_{RO}, \tag{10}$$

The last equation expresses the ideal system case, which means there is no noise affecting the system. From which:

$$\phi = \cos^{-1} \sqrt[k]{\frac{P_R}{P_{RO}}}, \tag{11}$$

where $k = m + 3$. In the previous works, P_{Ro} is known while here, P_{Ro} is not known exactly with noise power P_n. Assuming constant noise power in the room then the noise add to both the receiver power and irradiance angle. Neglecting the effect of NLoS (as it is very small), include the noise effect to (10) as follows:

$$P_R + P_n = \cos^{m+3}(\phi + \Delta\phi)(P_{RO} + P_n), \tag{12}$$

where P_n and $\Delta\phi$ are the receiver power and irradiance angle noises, respectively. Substituting ϕ from (11):

$$\Delta\phi = \cos^{-1} \sqrt[k]{\frac{P_R + P_n}{P_{RO} + P_n}} - \cos^{-1} \sqrt[k]{\frac{P_R}{P_{RO}}}. \tag{13}$$

From **Figure 1**, the horizontal distance without any noises is given by

$$d_L = V \tan(\phi). \tag{14}$$

In the case of a noisy channel, the horizontal-distance error Δd_L is estimated from:

$$d_L + \Delta d_L = V \tan(\phi + \Delta\phi). \tag{15}$$

The value of the horizontal-distance error Δd_L is used to determine the localization more accurate after calculating the position of the receiver by trilateration method as shown in the next section.

According to the RSS method, the positioning error is simply obtained from the distance errors. The positioning algorithm uses three maximum power levels to determine the location of the user [3]. Here, RSS algorithm is used to estimate (x_0, y_0), the position coordinates of the receiver. Let $(x_i, y_i), i \in \{1, 2, 3\}$, be the three coordinates of three transmitters. From Δd_L find the positioning estimate:

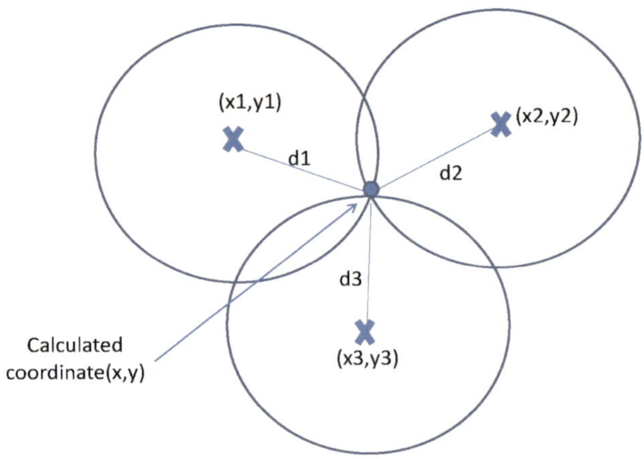

Figure 4.
A trilateration method for RSS to calculate the position of a receiver by using three transmitters.

$$(x_0 - x_i)^2 + (y_0 - y_i)^2 = \hat{d}_{L,i}^2, \tag{16}$$

where

$$\hat{d}_{L,i}^2 = d_L + \Delta d_L, \tag{17}$$

for any $i \in \{1, 2, 3\}$, where $d_{L,i}$ is the horizontal distance of the receiver from the transmitter i. This method can be clear as shown in **Figure 4**, where RSSI is received signal strength indicator.

4. Proposed KF algorithm

In this section, a KF algorithm is proposed to further improve the previous estimation (introduced in the last section) of the receiver position. Specifically, the estimation of the irradiance angle developed in the last section is further improved by adopting a KF algorithm.

The proposed system with KF is shown in **Figure 5** where the PD collects the received power and inserts it into the proposed system. The process of the proposed system is analyzing the mathematical equations to calculate the irradiance angle ϕ and the error of the irradiance angle $\Delta\phi$. Both of the two calculated values insert into the Kalman filtering process. The optimal value of the estimated irradiance angle is obtained after using the KF algorithm. The estimated irradiance angle inserted into the RSS process to calculate the estimated horizontal distance $\hat{d}$. The trilateration method has been applied to calculate the estimated position of the receiver by using the estimated horizontal distance $\hat{d}$. KF algorithm recursively

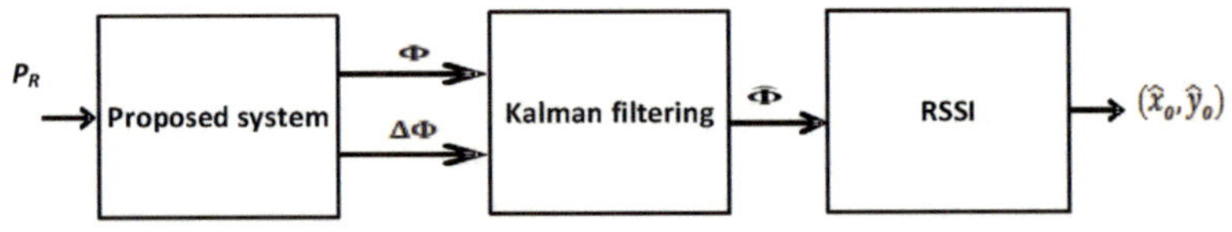

Figure 5.
The block diagram of the proposed system with KF.

estimates the state of variables in the system in two phases; prediction and measurement [4, 5].

4.1 Predict step

We denote the state vector by $x = (x_a, x_b)^T$, where x_a represents a measured angle, x_b is the error in the angle, and T is the transpose operator. Based on the estimate at iteration $k-1$, the state $x_{k-1|k-1}$. The next step k of the system dynamics $x_{k|k-1}$ is evaluated as:

$$x_{k|k-1} = F_k x_{k-1|k-1}, \tag{18}$$

where F_k is the state transition matrix. The corresponding state covariance matrix is given by:

$$P_{k|k-1} = F_k P_{k-1|k-1} F_k^T + Q_k, \tag{19}$$

where Q_k is the process noise covariance.

4.2 Measurement step

The updated state variable $x_{k|k}$ and updated state covariance matrix $P_{k|k}$ are given by

$$\begin{aligned} x_{k|k} &= x_{k|k-1} + K_k y_k \\ P_{k|k} &= (I - K_k H_k) P_{k|k-1}, \end{aligned} \tag{20}$$

respectively, where K_k is the Kalman gain, y_k is is the error vector, and H_k is is the observation model.

$$\begin{aligned} K_k &= P_{k|k-1} H_k^T S_k^{-1} \\ y_k &= z_k - H_k x_{k|k-1}. \end{aligned} \tag{21}$$

Here z denotes the measurement vector, given by:

$$z_k = H_k x_k + R_k \tag{22}$$

where R_k is the measurement noise matrix. Also S_k is the innovation matrix, which relates the covariance of state variables to measurement vector:

$$S_k = H_k P_{k|k-1} H_k^T + R_k. \tag{23}$$

Finally, after getting the estimated angle then recalculate the positioning error using equations developed in Section 3.

5. Proposed localization methodology using an averaging RSS technique

Second technique contains the averaging localization method and Kalman filtering with averaging schemes. For the averaging technique, the position of the receiver has been estimated by RSSI technique for multiple times (e.g., **N** samples) and the acquired estimations are averaged over all samples.

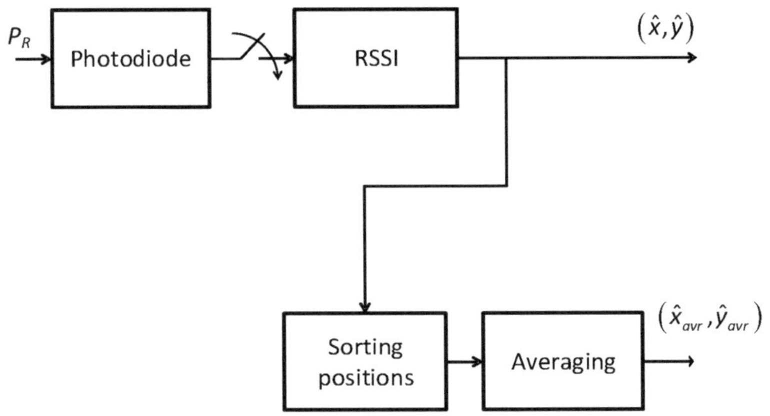

Figure 6.
Block diagram of proposed averaging positioning scheme.

To enhance the results of improving the localization, The algorithm of Kalman filtering has been adopted for estimation the received power over **N** samples, followed by using RSS technique on the average received power. These techniques have been analyzed mathematically, with respect to the effects of both LoS and first-reflection from NLoS propagation.

Typical room is considered for evaluating the positioning performances for proposed techniques and the results of them are compared with the traditional RSS system.

For determining the receiver location, the trilateration method is used with the RSS from three LEDs transmitters having the maximum received levels [3]. Our techniques depend on the average of estimated receiver position over a certain number of measurements to decrease the localization error. This decreasing in error gets at the cost of exceeding the system mathematical complexity. **Figure 6** shows a simple block diagram that demonstrates this approach.

5.1 RSS technique

Using (2), the received LoS power from transmitter $i \in \{1, 2, 3, 4\}$ can be written as:

$$P_{R,i} = \left(\frac{m+1}{2\pi d_i^2}\cos^{m+1}(\phi_i)A_R\right)P_{T,i}, \tag{24}$$

where $P_{T,i}$ is the transmitted power of i th LED. Here, assume that $\psi_i = \phi_i$, which is determined from **Figure 1** as:

$$\cos(\phi_i) = \frac{V}{d_i}, \tag{25}$$

where V is the vertical distance between transmitter and receiver, assumed constant. Accordingly, the distance between transmitter i and receiver can be evaluated as:

$$d_i = \sqrt[m+3]{\frac{(m+1)V^{m+1}A_R P_{T,i}}{2\pi P_{R,i}}}. \tag{26}$$

If consider the effect of NLOS as well, the total power collected at the receiver is obtained by modifying (24) to:

$$P_{R,i} = \left(H^{i}_{\mathrm{LOS}} + H^{i}_{\mathrm{NLOS}}\right) P_{T,i}. \tag{27}$$

5.2 Linear LS method

To estimate the receiver location, the linear LS estimation is commonly used. Let $(x_i, y_i), i \in \{1, 2, 3\}$, be the horizontal coordinates of transmitter i and $d_{L,i}$ be the horizontal distance of the receiver from transmitter i. The range equation can be written in the form:

$$(\hat{x} - x_i)^2 + (\hat{y} - y_i)^2 = d^2_{L,i}, \qquad i \in \{1, 2, 3\}, \tag{28}$$

where $(\hat{x}, \hat{y})$ is the estimated horizontal location of receiver. The last system of equations can be written in matrix form as:

$$A\hat{X} = B, \tag{29}$$

where

$$\begin{aligned} \hat{X} &= [\hat{x} \quad \hat{y}]^T \\ A &= \begin{bmatrix} x_2 - x_1 & y_2 - y_1 \\ x_3 - x_1 & y_3 - y_1 \end{bmatrix} \\ B &= [b_{21} \quad b_{31}]^T. \end{aligned} \tag{30}$$

Here for any $m \in \{2, 3\}$,

$$b_{m1} = (\hat{x} - x_1)(x_m - x_1) + (\hat{y} - y_1)(y_m - y_1). \tag{31}$$

The solution of (29) is:

$$\hat{X} = \left(A^T A\right)^{-1} A^T B. \tag{32}$$

5.3 Complexity analysis

The complexity of proposed averaging RSS technique can be analyzed by counting the number of mathematical operations required to solve the LS method once and then multiplying the resulting by the number of samples. Specifically, the total number of floating-point operations is $39N + 1$ flops, where N is the number of samples. That is, the complexity increases linearly with the number of samples.

6. Kalman filtering with averaging

KF estimates the states of a linear system from the noisy measurements then produces the estimation of unknown variables that aim to get more accurate than those which based on a single measurement value.

At this section, a KF algorithm is adapted to enhance the estimation performance of the receiver positioning system. In the first, KF estimates several samples of measured received powers. Then, the average of these estimated power values is

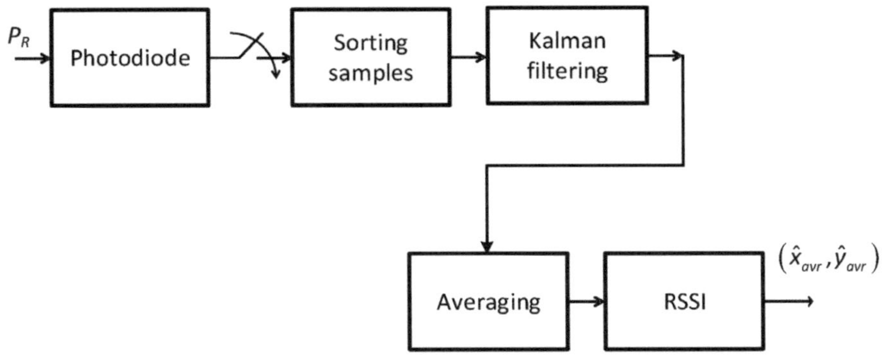

Figure 7.
Block diagram of proposed Kalman filtering technique.

evaluated. Using the output of KF which is the estimated average power, the position of the receiver can be calculated by using RSS technique. The block diagram of the proposed Kalman filtering with averaging technique is shown in **Figure 7**.

KF algorithm is shown in the previous chapter in Section 4 recursively estimates the state of variables in the system in two phases; prediction and measurement [4, 5]. We denote the state vector by x. This state vector represents measured received power and number of samples which are used in the process. Based on the estimate at iteration $k - 1$, and have state $x_{k-1|k-1}$.

7. Simulation and discussion for averaging technique

In this section, The simulation results are presented and compared them to that of traditional systems. **Table 1** shows the main parameters used in the simulation.

In case of demonstrating the relation between the SNR and the average positioning error, **Figure 8** shows that using five different positions of the receiver and the average positioning error in a meter. This figure shows that the proposed system outperform the traditional RSS method by nearly 1 cm at SNR=10 dB while adopting KF decreases the error by 11.5 cm that means improvement by 52.27%.

Figures 9 and **10**, plot the true path with three different methods. The simulation is done at a FoV of 70° and an SNR of 20 dB. These figures are simulated as a plan view to show the estimation position of the receiver in the room. The proposed RSS technique achieves a tiny improvement while the proposed RSS with KF is the closest to the true path.

Parameter	Value
Room dimensions	$5 \times 5 \times 3\ m^3$
Number of transmitters	4
Transmitted power	30 W
Locations of LEDs	(1.25, 1.25, 3), (1.25, 3.75, 3), (3.75, 1.25, 3), (3.75, 3.75, 3) m
FoV of photodetector	{70°}
The active area of the photodetector	$1\ cm^2$

Table 1.
Main parameters in VLC.

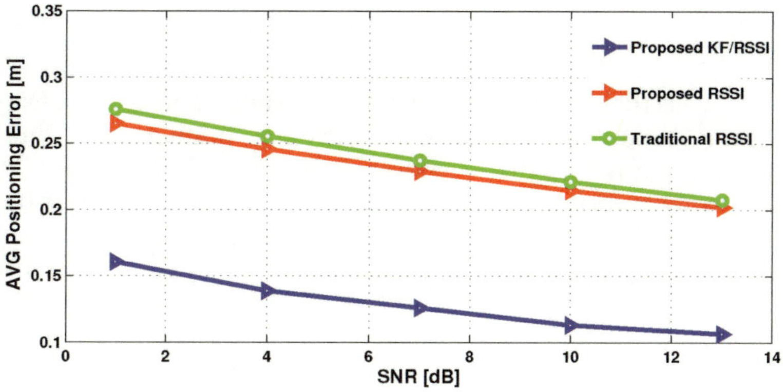

Figure 8.
The relation between average positioning error in meter with SNR for different positions.

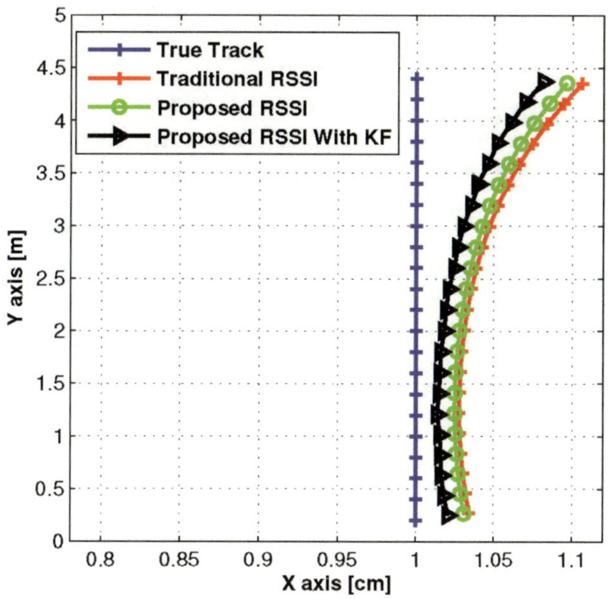

Figure 9.
Positioning error for traditional proposed, and KF correction RSS techniques at a FOV of 70° and an SNR of 20 dB for a y path.

The pedestrian is moving in random directions inside the room. In **Figure 11**, compare between three techniques: Traditional RSS, proposed RSS, and proposed RSS with Kalman filtering. The comparison is done at a FoV of 70° and an SNR of 20 dB. It is clear from the figure that adopting KF estimation further reduces the positioning error and provides an estimate that is very close to reality.

In the second technique, simulation results for the proposed system are presented and compared with that of traditional systems. The main parameters used in the simulations for the VLC link are listed in **Table 2**.

7.1 Positioning error

In the simulation, the performance measure is determined by the positioning error:

$$\varepsilon_{\text{position}} = \sqrt{(\hat{x} - x_0)^2 + (\hat{y} - y_0)^2}, \tag{33}$$

where (x_0, y_0) is the receiver horizontal location and $(\hat{x}, \hat{y})$ is its estimated location. **Figure 12** shows the average error in receiver positioning for different number of samples. It is clear that the error can be reduced to less than 10% of its maximum value by averaging over 50 samples. This reduction comes at the cost of

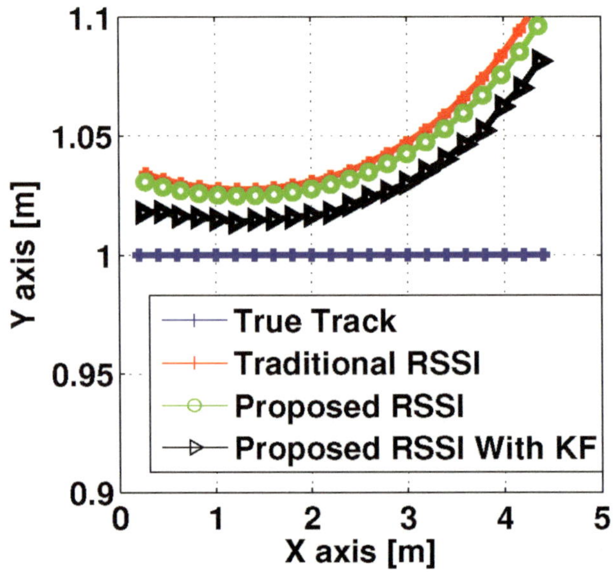

Figure 10.
Positioning error for traditional, proposed, and KF correction RSS techniques at FOV of 70° and an SNR of 20 dB for a x path.

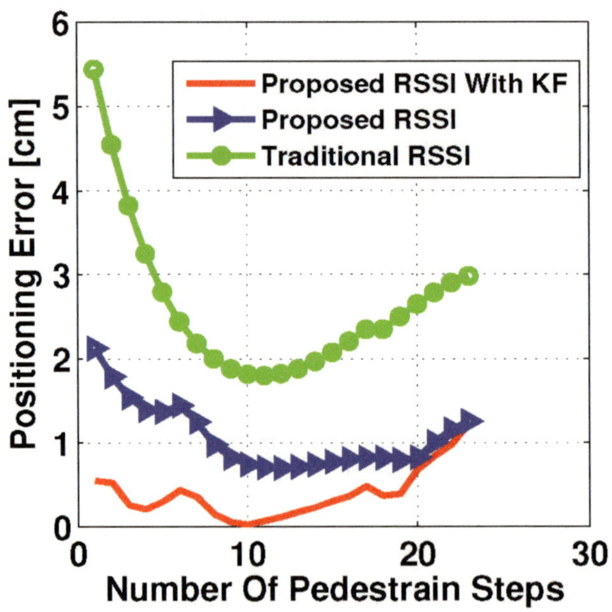

Figure 11.
Relation between positioning error and number of pedestrian steps at a FOV of 70° and an SNR of 20 dB.

Parameter	Value
Room dimensions	$5 \times 5 \times 3$ m^3
Number of transmitters	4
Total transmitted power	30 W
Locations of LEDs	(1.25, 1.25, 3), (1.25, 3.75, 3), (3.75, 1.25, 3), (3.75, 3.75, 3) m
FOV of photodetector	70°
SNR	20
Active area of photodetector	1 cm^2
Wall reflectivity ρ	0.8
Number of samples	50
Range of receiver in room	(1–3.5) m over both x, y axes

Table 2.
Simulation parameters.

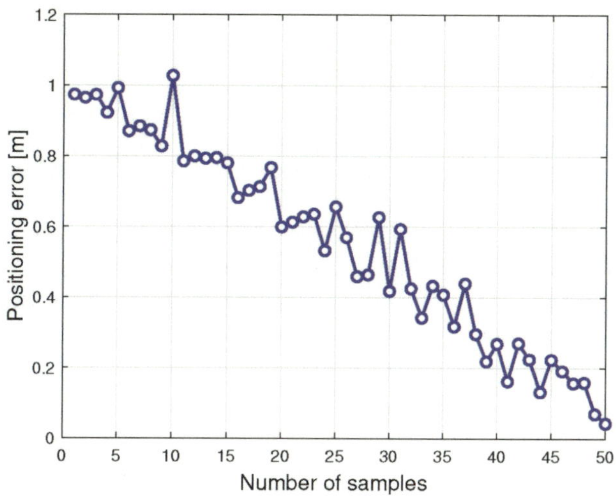

Figure 12.
Comparison between average positioning errors versus number of samples in averaging RSS technique.

increasing the mathematical complexity of the system as shown in **Figure 13**. The complexity is calculated according to the number of operations, which increases as the number of samples increases.

7.2 Averaging RSS and traditional RSS techniques

The RSS variations of the positioning error at every sample are shown in **Figure 13** for receiver position $(x_0, y_0) = (1, 1)$, considering the effect of LoS only.

The positioning error using the proposed averaging RSS technique (with 100 samples) is plotted in same figure as well. The improvement using proposed technique is clear from the figure. The traditional RSS errors are more than 0.6 m (42.4%), where the error when employing the proposed averaging RSS is only 0.217 m (15.3%). That is, an improvement of about 27.1% is getting when adapting the proposed system. Both LoS and NLoS effects are studied for position of the receiver $(x_0, y_0) = (1, 1)$ as well and the results are plotted in **Figure 14**. Traditional RSS errors are more than 0.7 m (49.5%), where the errors when using proposed

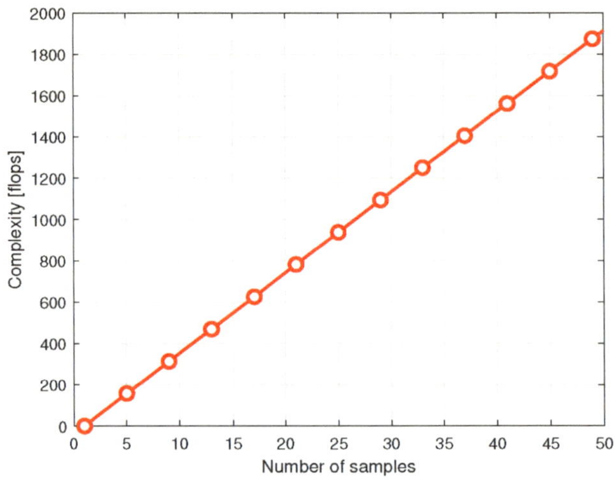

Figure 13.
Complexity of the averaging RSS method.

averaging RSS are only 0.255 m (18%). The improvement of nearly 31.5% is achievable by using the proposed scheme.

8. Kalman filtering, averaging RSS, and traditional RSS techniques

In this section, different comparisons are shown between the performance of three methods; The traditional RSS technique, proposed averaging RSS technique, and the proposed Kalman filtering with averaging. We use same values which given for the VLC link of **Table 2**.

8.1 LoS propagation

The effects of LoS only on two tracks' estimations for both x and y paths are presented in **Figures 14** and **15** for both x and y paths, respectively.

From the figures, both tracksâ€™ estimations are the nearest to the real one when employing the proposed techniques. Also, The results show that adopting KF estimation reduces the positioning error and improve the estimation. **Table 3** for three techniques summarizes the error and improving percentages.

8.2 Both LoS and NLoS propagations

The effect of both LOS and NLOS on Kalman filtering tracks' estimations for both x and y paths are presented in **Figures 16** and **17** for both x and y paths, respectively. It is clear that the track estimation gets slightly worse when including the effect of NLoS.

8.3 Kalman filtering response

The response for a random position estimation for the KF is shown in **Figure 18**. The filter input is a measured value of received power, while the filter output is the corresponding estimated value at different number of samples. The filter response (estimated value) is near to the actual value where the samples are greater than 11.

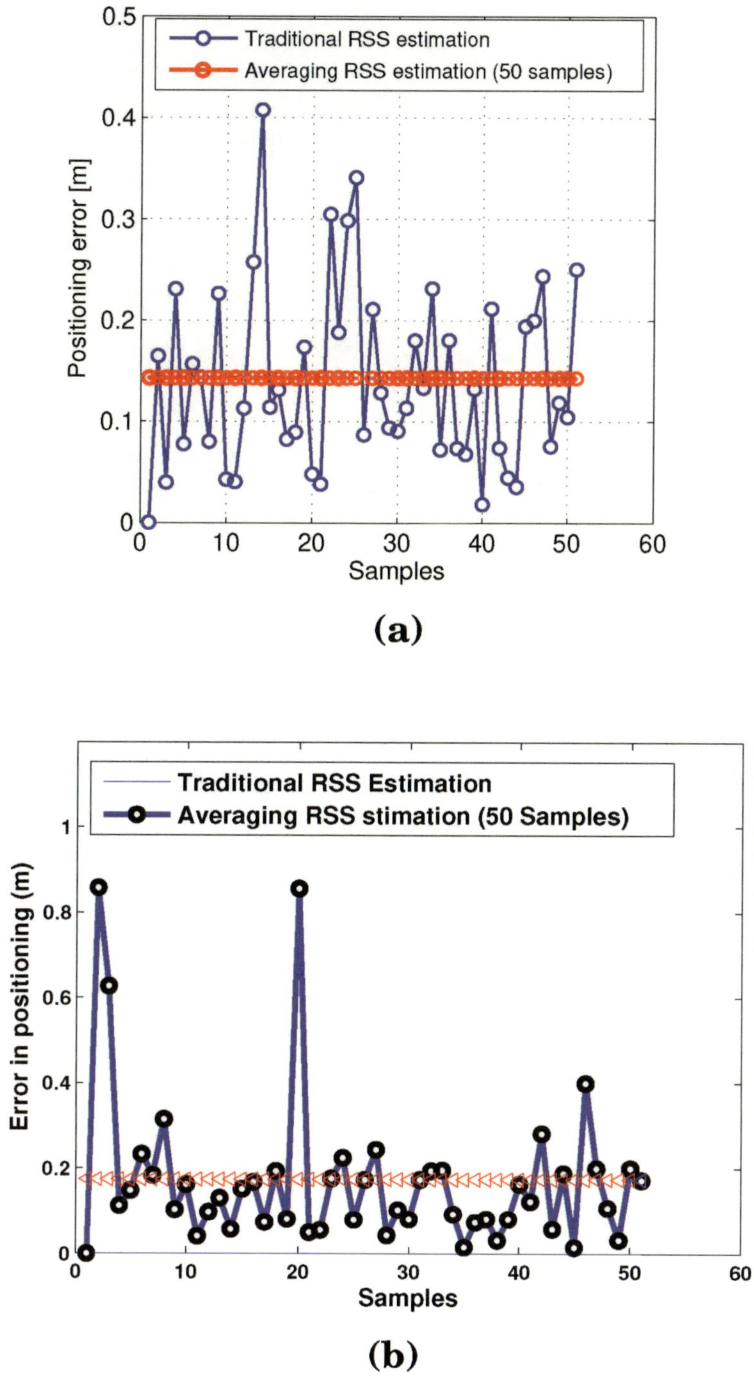

Figure 14.
Positioning error for both traditional RSS and averaging RSS techniques at position (1, 1), *considering the effects: (a) LOS only, (b) both LOS and NLOS.*

8.4 Position estimation accuracy comparison

As mentioned in the introduction, several techniques have been proposed for indoor localization based on VLC technology. In this section, a comparison is

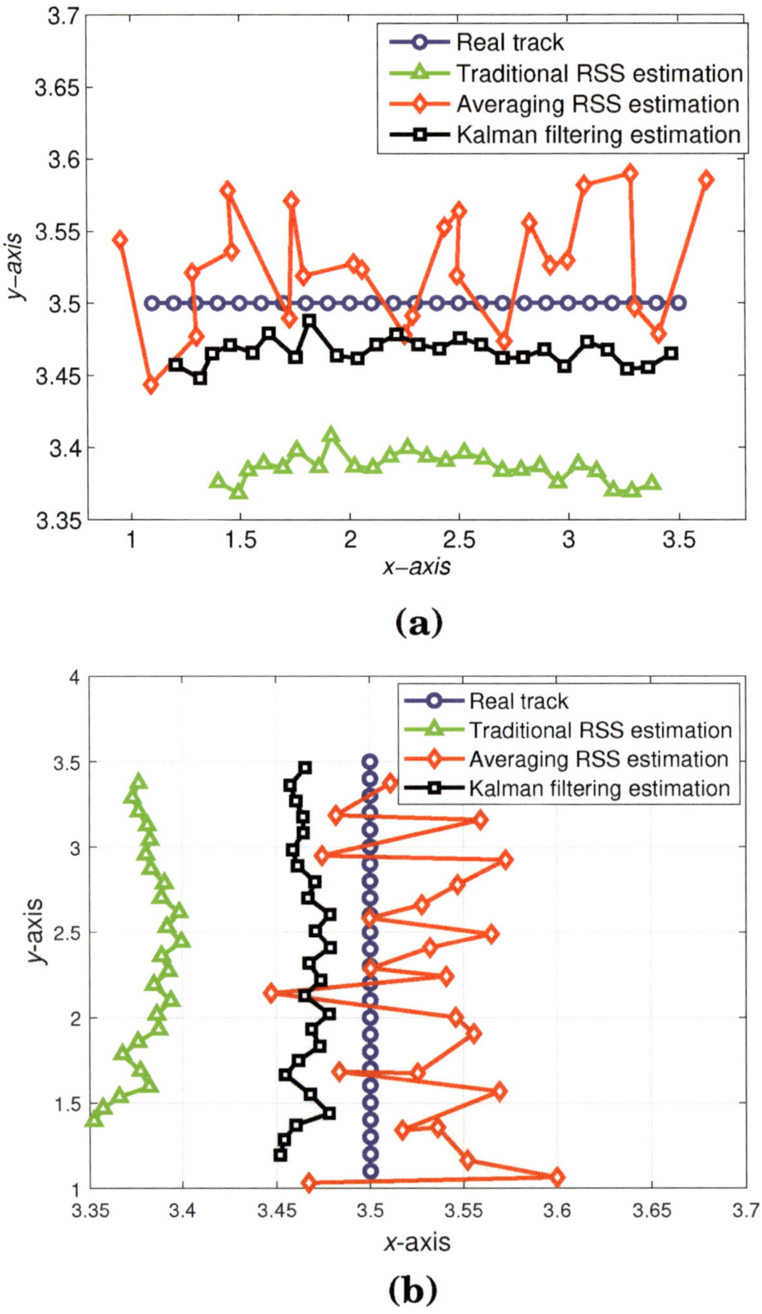

Figure 15.
Track estimations using traditional and proposed techniques for: (a) an x path, (b) an y path.

Localization method	Average positioning error	Percentage improvement
Traditional RSS	18 cm	—
Averaging RSS	12 cm	33.3%
Kalman filtering	5 cm	72.2%

Table 3.
Accuracy for different techniques.

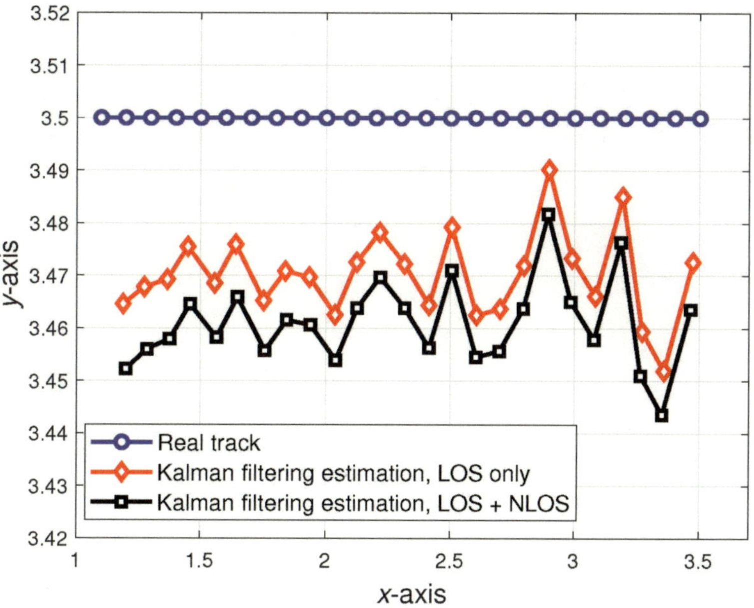

Figure 16.
Comparison between Kalman filtering track estimation for both LOS and NLOS propagations for an x path.

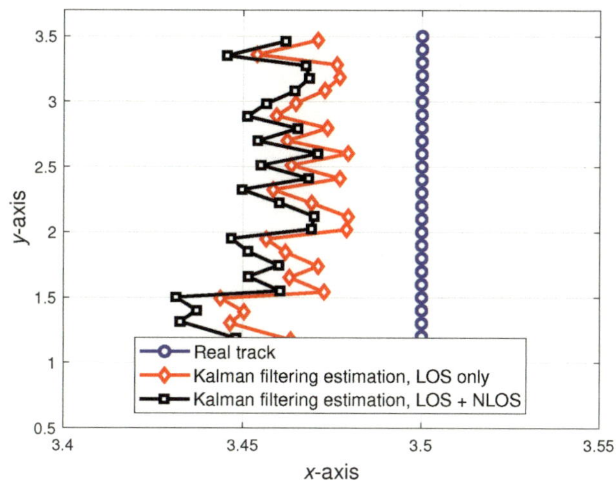

Figure 17.
Comparison between Kalman filtering track estimation for both LOS and NLOS propagations for an y path.

provided between the position estimation accuracy and that of previous works for same simulation parameters. The results of this comparison are summarized in **Table 4**.

It is clear from the **Table 3** that both proposed averaging RSS and Kalman filtering with averaging techniques achieve better accuracy than that proposed in [6–8]. Since the authors in [9, 10] have adopted Kalman filtering, they have better accuracy than the proposed averaging method. However, employing Kalman filtering with averaging gives a better accuracy.

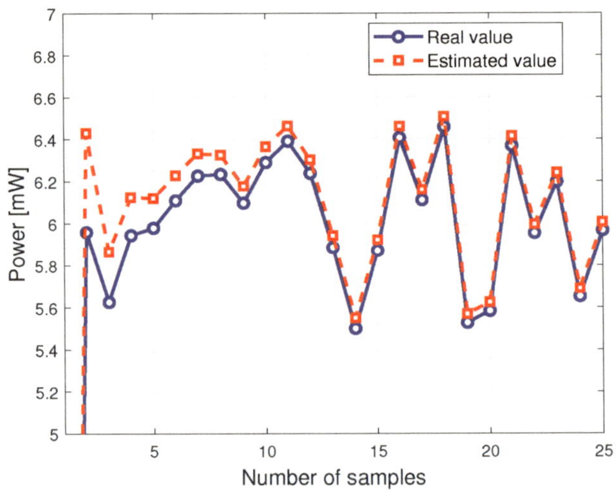

Figure 18.
Response of Kalman filtering technique.

Reference	Sys. parameters	Ref. acc.	Present work AVG sys. Acc.	Present work KF acc.
[6]	LoS, FoV = 80°, $P_T = 10$ W, $A_R = 0.5$ cm^2, $(5, 5, 3)$ m^3	5 cm	3.7 cm	3.1 cm
[7]	LoS/NLoS, FoV = 10–180°, $P_T = 1.9$ W, $A_R = 0.81$ cm^2, $\rho = 60\%$, $(4, 4, 3.5)$ m^3	13.95 cm	9.1 cm	4.8 cm
[8]	LoS, FoV = 85°, $P_T = 1$ W, $A_R = 0.81$ cm^2, $(5, 4, 3)$ m^3	10 cm	6.17 cm	1.75 cm
[9]	LoS, FoV = 80°, $P_T = 17$ W, $A_R = 1$ cm^2, $(3.6, 3.26, 2.5)$ m^3	14.5 cm	17.4 cm	3.5 cm
[10]	LoS, FoV = 25°, $P_T = 17$ W, $A_R = 1$ cm^2, $(6, 6, 3)$ m^3	5 cm	11 cm	2.3 cm

Table 4.
Position estimation accuracy comparison.

9. Concluding remarks

First, the proposed techniques have been analyzed mathematically, taking into account the effects of LoS propagation. The positioning estimation accuracy of proposed techniques has been evaluated in a typical room. The results reveal that an improvement of about 52% in the average positioning error is achievable using the proposed technique with KF, when compared to that of the traditional RSS.

Secondly, both averaging and Kalman filtering by averaging schemes are adapted to improve the positioning system. Specifically, in the averaging technique, the receiver position has been determined by using the average of the samples of RSS estimations. The position is determined by RSS estimation of a Kalman filtered averaged multiple received power samples in the second proposed system, Kalman filtering with averaging algorithm.

Simulation results reveal that an improvement of about 33.3% in estimation accuracy is achievable when using the averaging scheme as compared to that of traditional RSS scheme. This improvement increases to 72.2% when adopting proposed Kalman filtering with averaging scheme.

Author details

Eman Shawky Abd El-Fattah Amer
Alex University, Alexandria, Egypt

*Address all correspondence to: eman.shawky@alexu.edu.eg

References

[1] Ghassemlooy, Z. Popoola, and W. Rajbhandari. Optical wireless communications system and channel modelling with MATLAB. Boca Raton, 2nd Edition, 2013.

[2] C. Huang and X. Zhang. LOS-NLOS identification algorithm for indoor visible light positioning system. pages 575–578, Bali, Indonesia, Dec. 17–20, 2017.

[3] Maxim Shchekotov. Indoor localization method based on wi-fi trilateration technique. In Proc. 16th Conference of Fruct Association (ACP 2018), Pages 177–179, Oulu, Finland, Oct. 27–31, 2014.

[4] Greg Welch and Gary Bishop. An introduction to the Kalman filter. Technical Report 95-041, University of North Carolina at Chapel Hill, Chapel Hill, NC, USA, July 24, 2006.

[5] Yuta Teruyama and Takashi Watanabe. Effectiveness of variable-gain kalman filter based on angle error calculated from acceleration signals in lower limb angle measurement with inertial sensors. Computational and Mathematical Methods in Medicine, 10 (1155):398042(1–12), 2013.

[6] F. Mousa, N. Almaadeed, K. Busawon, A. Bouridane, R. Binns, and I. Elliot. Indoor visible light communication localization system utilizing received signal strength indication technique and trilateration method. Optical Engineering, 57(1): 016107(1–10), Jan. 2018.

[7] G. B. Prince and T. D. C. Little. A two phase hybrid RSS/AoA algorithm for indoor device localization using visible light. In IEEE Global Communications Conference (GLOBECOM 2012), Pages 3347–3352, Anaheim, CA, Dec. 3–7, 2012.

[8] A. Şahin, Y. S. Eroğlu, İ Güvenç, N. Pala, and M. Yüksel. Hybrid 3-D localization for visible light communication systems. Journal of Lightwave Technology, 33(22):4589–4599, Nov. 2015.

[9] Lihui Feng Zhitian Li and Aiying Yang. Fusion based on visible light positioning and inertial navigation using extended kalman filters. IEEE SENSORS, 17(1093):093â€“1103, JUNE 2017.

[10] Fatih Erden Yusuf Said Eroglu and Ismail Guvenc. Adaptive kalman tracking for indoor visible light positioning. Eess.SP, 1, Sep 2019.